Advances in Industrial Control

Advances in Industrial Control is a series of monographs and contributed titles focusing on the applications of advanced and novel control methods within applied settings. This series has worldwide distribution to engineers, researchers and libraries.

The series promotes the exchange of information between academia and industry, to which end the books all demonstrate some theoretical aspect of an advanced or new control method and show how it can be applied either in a pilot plant or in some real industrial situation. The books are distinguished by the combination of the type of theory used and the type of application exemplified. Note that "industrial" here has a very broad interpretation; it applies not merely to the processes employed in industrial plants but to systems such as avionics and automotive brakes and drivetrain. This series complements the theoretical and more mathematical approach of Communications and Control Engineering.

Indexed by SCOPUS and Engineering Index.

Publishing Ethics

Researchers should conduct their research from research proposal to publication in line with best practices and codes of conduct of relevant professional bodies and/or national and international regulatory bodies. For more details on individual ethics matters please see:
https://www.springer.com/gp/authors-editors/journal-author/journal-author-helpdesk/publishing-ethics/14214

More information about this series at http://www.springer.com/series/1412

Ai Hui Tan · Keith Richard Godfrey

Industrial Process Identification

Perturbation Signal Design and Applications

Springer

Ai Hui Tan
Faculty of Engineering
Multimedia University
Cyberjaya, Selangor, Malaysia

Keith Richard Godfrey
School of Engineering
University of Warwick
Coventry, Warwickshire, UK

ISSN 1430-9491 ISSN 2193-1577 (electronic)
Advances in Industrial Control
ISBN 978-3-030-03660-7 ISBN 978-3-030-03661-4 (eBook)
https://doi.org/10.1007/978-3-030-03661-4

Library of Congress Control Number: 2018961195

This Springer imprint is published by the registered company Springer Nature Switzerland AG
The registered company address is: Gewerbestrasse 11, 6330 Cham, Switzerland

To my parents
Ai Hui Tan

Series Editors' Foreword

The series *Advances in Industrial Control* aims to report and encourage technology transfer in control engineering. The rapid development of control technology has an impact on all areas of the control discipline. New theory, new controllers, actuators, sensors, new industrial processes, computer methods, new applications, new design philosophies, new challenges, etc. Much of this development work resides in industrial reports, feasibility study papers and the reports of advanced collaborative projects. The series offers an opportunity for researchers to present an extended exposition of such new work in all aspects of industrial control for wider and rapid dissemination.

Parameter estimation techniques for static and dynamical systems were probably given a first thorough theoretical basis in the years around 1800. In about 1795, Gauss claimed to have already developed and used the least squares estimation method, a method that was given in definitive published form in 1805, but by Legendre. Gauss's own publication of the method he was using came a little later in 1809. All this initiated one of those scientific priority claim issues that is still discussed today (See *Gauss and the Invention of Least Squares* by S. M. Stigler in *The Annals of Statistics*, 1981, Vol. 9, No. 3, pp. 465–474).

Approximately one hundred and sixty years later, least squares came to be an important tool in system identification for control systems design methods. However, we like to think that before that, system identification for control design began with the nonparametric methods of step responses and the 'sustained oscillation' experimental methods of Ziegler and Nichols in the 1940s. In the early 1980s, these methods were given renewed relevance with the elegant 'relay experiment' method of Åström and Hägglund. The *Advances in Industrial Control* monograph series has many volumes on the related design of PID controllers and a number of volumes dealing with the relay experiment paradigm, including:

- *Autotuning of PID Controllers* by Cheng-Ching Yu (ISBN 978-3-540-76250-8, 1999 [2nd edition, published as a stand-alone title: ISBN 978-1-84628-036-8, 2006]);
- *Industrial Process Identification and Control Design* by Tao Liu and Furong Gao (ISBN 978-0-85729-976-5, 2012); and

- *Non-Parametric Tuning of PID Controllers* by Igor Boiko (ISBN 978-1-4471-4464-9, 2012).

The advent of state-space system formalism for linear and nonlinear models completely changed the landscape of control system theory. The work of Rudolf Kalman and co-authors published in a key set of papers in the 1960s initiated a different approach to systems theory and control systems design that was able to subsume both time domain and frequency domain descriptions of dynamical processes. The use of state-space system descriptions gave a significant impetus to system identification methods for estimating the parameters of state-space and transfer function models. The least squares algorithm adapted to both batch and recursive computations became a fundamental tool in this new paradigm. The sophisticated methods of system identification now include techniques such as prediction error identification and subspace identification methods. This level of theoretical maturity requires specialised authors to present the new ideas and relevant *Advances in Industrial Control* monograph series and the *Advanced Textbooks in Control and Signal Processing* series offerings include:

- *Practical Grey-box Process Identification* by Torsten Bohlin (ISBN 978-1-84628-402-1, 2006); and
- *Identification of Continuous-time Models from Sampled Data* edited by Hugues Garnier and Liuping Wang (ISBN 978-1-84800-160-2, 2008); and
- *System Identification* by Karel J. Keesman (ISBN 978-0-85729-521-7, 2011).

Practical model identification for industrial systems can either use straightforward plant operational records or use an additional injected perturbation signal. Operational data may not give very good estimation efficiency so the injected identification signal is often the better strategy. But here operational constraints may forbid any injected signal amplitude that causes observable end-product quality degradation. Further, the identification signal should be designed to maximise estimation efficiency. This *Advances in Industrial Control* monograph entitled *Industrial Process Identification: Perturbation Signal Design and Applications* by authors: Ai Hui Tan and Keith Richard Godfrey reports their research on these very issues. They design pseudorandom binary signals to avoid disturbing the process performance significantly yet seek to optimise estimation accuracy. The monograph has a good balance between theory and applications. Academic examples illustrate the theory and a major attraction of the monograph is the inclusion of two case studies. Chapter 6 reports the modelling of an electronic nose system with direction-dependent properties. Chapter 7 presents an application to a multivariable cooling system; from the process control domain. The monograph closes with Chap. 8 that gives assessments and advice for using standard widely available software in this field.

Glasgow, Scotland, UK

Michael J. Grimble
M. A. Johnson
Industrial Control Centre
University of Strathclyde

Preface

Perturbation (input) signal design plays a very important role in system identification for control applications. A good design can lead to maximally informative experiments which reduce operational costs associated with identification tests. Many different approaches to the design of perturbation signals have been described over the years. However, the selection of the most suitable signal for a particular industrial process application remains a challenging task. This monograph helps the reader to understand the different designs that are particularly relevant in an industrial context and guides the reader to make well-informed choices. The major aspects of perturbation signal design such as the formulation of suitable specifications in the face of practical constraints, the classes of designs available, the various objectives necessitating separate treatments when dealing with nonlinear systems and extension to multi-input scenarios are considered. Two Case Studies are included to provide a good balance between theory and practice. Several software packages which are readily available on the Web are discussed, so that readers can access designs for their particular application and do not have to go through the time-consuming procedure of designing signals for themselves.

In the measurement of the dynamics of a system, a perturbation signal is applied to the input of the system, and the response to it is measured at the output. These input and output signals are then processed to give an estimate of the system dynamics. This may be in the form of either a nonparametric estimate or a parametric estimate. Examples of nonparametric estimates include impulse responses or step responses in the time domain or frequency responses in the frequency domain, while examples of parametric estimates include continuous or discrete transfer functions and state-space descriptions.

In practice, there are many important questions about the design of perturbation signals that need to be considered. Among these are the following:

- If the system is noisy, then to obtain sufficiently accurate estimates of the system dynamics, it is necessary either to average over a long period of time (which may not be feasible in industrial systems) or to increase the perturbation signal amplitude, which means that any assumption of linearity may no longer be

valid. Thus, are there signals that enable the effects of any nonlinearities to be minimised, so obtaining a good linear model of the system?
- Is it possible to design perturbation signals that are particularly suitable for characterising various aspects of the nonlinear behaviour itself?
- What are the possible signal designs for the general multi-input scenario as well as for cases where the process is ill-conditioned?

This book is intended for industrial engineers especially control engineers, and for researchers in process control and communication engineering; also for academics wishing to learn about the most recent results in perturbation signal design and final year undergraduate and postgraduate students in electrical, mechanical and communication engineering wishing to learn and apply perturbation signal design in project work. It will be particularly helpful to those seeking guidance on choosing identification software tools for use in practical experiments and Case Studies.

The three most important aspects of this book are:

- Several software packages are discussed and explained so that the reader finds it easy to get started on designing some common classes of signals.
- Experiments on real systems are discussed which provide readers with insights on how to design suitable perturbation signals in practical scenarios.
- There is a lack of books on perturbation signal design and such a book has been greatly overdue since many new designs have appeared in recent years.

There are eight chapters in this book. A summary of what is included in each of these follows.

Chapter 1 sets the scene for the rest of this book. It starts with a short review of early applications (in the 1960s) of system identification using periodic perturbation signals to full-scale industrial processes and nuclear power plants; more recent applications of identification techniques in industry feature in the remainder of this book. The role of signal design in identification for control is then expounded. Time domain and frequency domain specifications for the identification of linear systems are explained. Performance measures for linear system identification are described. The identification framework is then extended to nonlinear systems where the concept of harmonic suppression is explained. The chapter concludes with a comparison between periodic and nonperiodic signals.

Chapter 2 considers the design of pseudorandom signals for linear system identification; these have fixed autocorrelation functions, and therefore fixed spectra, dependent only on the class and period of the signal. The theory behind the generation of several classes of pseudorandom signals is discussed and their properties are explained. These include maximum length binary signals based on shift register sequences, as well as several other classes of binary and near-binary signals. The maximum length design extends to the multilevel case, where the sequence-to-signal conversion determines not only the harmonic properties but also the period of the resulting signal. An interesting class of truncated signals arises from certain well-defined choices. The design of direct synthesis ternary signals and the suboptimal direct synthesis ternary signals is discussed. The chapter concludes with an application example for linear identification of a system in the presence of nonlinear distortion.

Chapter 3 discusses the design of computer-optimised signals for linear system identification. Unlike pseudorandom signals which have fixed spectra, the objective now is to design a signal with a spectrum as close as possible to a specified spectrum. The optimisation algorithms come in different forms, depending on the class of signal, and particularly the number of signal levels. The classes of signals considered include multisine sum of harmonics signals which can take any value between their minimum and maximum, discrete interval signals which are either binary or ternary and multilevel multiharmonic signals which have a small number of signal levels, where this number is specified by the user. It is then shown that it is also possible to combine the advantages of pseudorandom and computer-optimised designs leading to a class of hybrid signals, which are generated as a combination of pseudorandom signals and computer-optimised ones. The concept of optimal input signals, where the power spectra are optimised based on initial models to satisfy an application-related objective, is briefly described.

Chapter 4 describes perturbation signal designs for multi-input system identification, giving the theory behind the design of sets of uncorrelated signals; these allow the effects of the individual inputs to be easily decoupled at the system output. The designs include Hadamard-modulated signals and signals with a zippered spectrum. The latter may be designed using multisine signals or pseudorandom signals. An application example on a simulated thermoelectric system is presented. An alternative approach using a phase-shifting design is described; here only one signal is generated and phase-shifted versions of this signal are applied to the multiple inputs. The identification of ill-conditioned processes is then discussed. The problem of ill-conditioning is caused by the singular values having widely differing magnitudes. The method of virtual transfer function between inputs for the identification of ill-conditioned systems is explained. Its effectiveness is illustrated on a simulated multizone furnace.

In Chap. 5, the identification setting is turned to systems with nonlinearities and time-varying properties. For systems with nonlinearities, the signal design is shown to be highly dependent on the objectives of the identification test. The identification of the best linear approximation is discussed next and is shown to be very useful when a nonlinear process is to be linearised around its operating point. While the best linear approximation depends on the perturbation signal applied, those with Gaussian amplitude distribution are advantageous particularly in the identification of block-oriented systems. For the measurement of Volterra kernels, it is shown that multisine signals with specially designed harmonics enable the kernels to be measured without interharmonic distortion. The chapter concludes with an exposition of a method based on frequency domain analysis which allows the quantification of the effects of nonlinearities, noise and time variation. The technique requires only a single experiment and in the case of multi-input systems, makes use of a set of uncorrelated perturbation signals. The effectiveness of the technique is illustrated on a mist reactor system.

Chapter 6 is the first of two Chapters devoted to a Case Study. In this example, the identification of an electronic nose system having direction-dependent properties, where the dynamics depend on whether the output is increasing or decreasing, is described. It is shown that the use of maximum length binary signals allows the detection of the nonlinearities through the input–output crosscorrelation function due to the shift-and-multiply property. When inverse-repeat maximum length binary signals are applied, the effects of even-order nonlinearities can be eliminated. The detection of the even-order nonlinearities is also possible through analysis of the output spectrum. The best linear approximation of the system is estimated using various methods in the time and frequency domains. It is also shown that the identification tests lead to the estimation of a Wiener model for the system and that the use of an inverse-repeat perturbation signal results in higher accuracy in the parameter estimates.

Chapter 7—in this second Case Study Chapter, the application of uncorrelated multisine signals for the identification of a multivariable cooling system with time-varying delay is illustrated. The system has two inputs, one associated with the flow control system and the other with the Peltier system, and a temperature output. The suppression of harmonic multiples of two and three in the signals confirms that the system is largely linear. The application of several periods of the perturbation signals leads to the identification of a time-invariant model for the channel from the Peltier system to the temperature output and the detection of a time-varying component for the channel from the flow control system to the temperature output. A delay reconciliation technique is applied to remove the relative delays between individual periods of the output so that the averaged period becomes more representative of the individual periods. Continuous-time modelling of the time-invariant part of the time-varying channel is further carried out as it facilitates online adaptive identification of the variable delay using a gridding approach which accommodates fractional delay values.

Chapter 8 discusses and explains several software packages for perturbation signal design so that the reader finds it easy to get started on designing some common classes of signals. These packages include *prs* (for maximum length binary and other binary and near-binary pseudorandom signals as well as direct synthesis ternary signals), GALOIS (for pseudorandom multilevel signals), functions in the Frequency Domain System Identification Toolbox (for multisine signals, discrete interval binary and ternary signals, and optimal input signals), *multilev_new* (for multilevel multiharmonic signals) and Input-Signal-Creator (for pseudorandom multilevel signals and uncorrelated signal sets formed from these).

We would like to thank Prof. H. A. Barker, Chin Leei Cham and Dr. Timothy Yap and for their excellent collaboration over many years; part of this collaborative work appears in this book. We would also like to acknowledge our many other collaborators, colleagues and friends for their suggestions, support and encouragement. We would like to express our gratitude to the Series Editors Prof. Michael J. Grimble and Prof. Michael A. Johnson, Springer Editor Oliver Jackson as well as the staff in Springer for making this book a reality. Last but not least, we would like to thank the IEEE, the IET and Elsevier for permission to reproduce copyrighted materials in this book.

Cyberjaya, Malaysia — Ai Hui Tan
Coventry, UK — Keith Richard Godfrey

Contents

Symbols, Operators and Abbreviations

Symbols

Only frequently occurring symbols and their main usage are listed.

A_p	Amplitude of harmonic p
c	Coefficients of feedback shift register
$f_{\text{resolution}}$	Frequency resolution
f_s	Sampling frequency
F	Number of specified excited harmonics
g	Primitive element of GF
G	FRF of a system, or
	Transfer function of a system
H	Hadamard matrix, or
	VTFBI
i	Discrete-time index
j	$\sqrt{-1}$
k	Harmonic number, or
	General index
M	Moment of a signal
M	Number of signal levels of MLMH signals, or
	Number of periods of a periodic signal, or
	Number of inputs of multivariable system
n	Discrete-time index, or
	Degree of feedback shift register
N	Period of a periodic signal
q	GF
R_{uu}	Autocorrelation function of the signal u
R_{uy}	Crosscorrelation function between u and y
$s_{q,n}$	PRML sequence from GF(q) with primitive polynomial of degree n
t	Continuous-time index

t_s	Sampling interval
T_N	Measurement period = $N \times t_s$
u	Input signal in the time domain
$U(k)$, $U(\mathrm{j}\omega)$	Input signal in the frequency domain
$U(s)$	Laplace transform of input signal
$U(z^{-1})$	z-transform of input signal
ω	Angular frequency
y	Output signal in the time domain
$Y(k)$, $Y(\mathrm{j}\omega)$	Output signal in the frequency domain
$Y(s)$	Laplace transform of output signal
$Y(z^{-1})$	z-transform of output signal
Z	Set containing integers
ϕ_p	Phase of harmonic p
σ	Singular value, or Standard deviation

Operators

$\mathbf{A}^T$	Matrix transpose
$\mathbf{A}^{-1}$	Matrix inverse
det	Determinant
E	Expectation
mod(v,w)	Modulo operator giving the remainder after the division of v by w
sup	Supremum
$\overline{U}$	Complex conjugate of U
var	Variance
$\oplus_q$	Modulo-q addition

Abbreviations

ARMAX	Autoregressive moving average with exogenous input
ARX	Autoregressive with exogenous input
CDMA	Code division multiple access
DFT	Discrete Fourier transform
DIB	Discrete interval binary
DIT	Discrete interval ternary
ELiS	Estimator for Linear Systems
EMINE	Effective minimum ratio between the actual amplitude and the specified amplitude at any of the specified harmonics
FRF	Frequency response function

GF	Galois field
GUI	Graphical user interface
HAB	Hall binary
MAE	Mean absolute error
MLB	Maximum length binary
MLMH	Multilevel multiharmonic
MSE	Mean square error
NID	No interharmonic distortion
PAPR	Peak-to-average power ratio
PIPS	Performance index for perturbation signals
PIPSE	PIPS (effective)
PRB	Pseudorandom binary
PRML	Pseudorandom multilevel
PWM	Pulse width modulation
QRB	Quadratic residue binary
QRT	Quadratic residue ternary
RGA	Relative gain array
RMS	Root mean square
SNR	Signal-to-noise ratio
TPB	Twin prime binary
VTFBI	Virtual transfer function between inputs
ZOH	Zero-order hold

Chapter 1
Introduction

1.1 Historical Perspective of Industrial Applications of System Identification

In the 1960s, there was considerable effort to improve the operation and control of existing industrial processes, and the design of new plant. This led to increasing interest in the measurement of system dynamics.

Time domain methods found increasing use, but step response and pulse response measurements often suffered from the drawback that, in the presence of noise, either the perturbation signal had to be so large that the normal operation of the process was affected, or the experimentation time needed to be so long that there were difficulties in holding the process steady over the necessary time.

This led to interest being focused on the use of pseudorandom signals applied at the input, and input–output crosscorrelation methods being used to obtain estimates of the impulse response of the system. A surprisingly large number of applications to full-scale industrial plant (including nuclear reactors) were reported in the literature during the 1960s. These were listed by Godfrey (1969a), with Table 3 in that paper listing eight applications of pseudorandom signals in the process industries (chemical, electricity supply, oil, paper and steel) and Table 4 listing similar applications to six different nuclear power systems. All applications except one used binary signals, the exception being the work of Chang et al. (1968), who used a five-level pseudorandom signal on a cold rolling mill at the Port Talbot Steelworks in Wales, with the objective of reducing the effects of nonlinearities on the (linear) impulse response estimates.

One of the examples of the application of the crosscorrelation technique using pseudorandom binary (PRB) signals to a full-scale industrial plant was conducted at the British Petroleum Oil Refinery in Belfast (Godfrey 1969b), as part of an agreement between BP and the National Physical Laboratory, UK. The process chosen was the 30-plate distillate splitter column with its sidestream stripper, which was part of the integrated crude oil distillation and distillate hydro-desulphurising unit at this

A. H. Tan and K. R. Godfrey, *Industrial Process Identification*, Advances in Industrial Control, https://doi.org/10.1007/978-3-030-03661-4_1

refinery. Three inputs (feed flow, reboiler temperature, and feed-heater temperature) were initially chosen on the grounds that they were the most relevant to normal operation of the column and they were perturbed separately using a 255-digit PRB signal. Three column temperatures were chosen as the output variables. The column and the rest of the refinery were operating normally under closed-loop control, and it was found that in this case, perturbing the feed flow had little effect on the output temperatures.

Measurements for the remaining six input–output pairs were analysed further. In each case, pulse responses were estimated and compared with corresponding pulse responses from simple step tests. The comparisons for two of the measured temperatures were very satisfactory, but one of the temperature measurements was very noisy and the crosscorrelation results proved much more satisfactory in this case. Discrete (z-domain) pulse transfer functions of the form

$$z^{-r}\frac{b_1 z^{-1} + b_2 z^{-2} + \cdots + b_k z^{-k}}{1 + a_1 z^{-1} + a_2 z^{-2} + \cdots + a_k z^{-k}} \tag{1.1}$$

were also estimated from the input–output data using a generalised least squares procedure. It was found that, with the relatively large sampling interval, an initial time delay of $r = 0$ gave the best results in terms of error between model and process output, while fourth-order models ($k = 4$) gave negligible improvement over the second-order models ($k = 2$). The main practical problems encountered in the work were that the perturbation signal amplitudes had to be kept very small so that the normal operation of the refinery was not disturbed, and that it was not easy to keep the unit completely stationary over long periods of time.

As mentioned earlier, more recent applications of identification techniques in industry will feature in the remainder of this book, as the techniques are described.

1.2 Role of Signal Design in Identification for Control

The selection of perturbation signals is an important step in the design of an experiment for system identification. A better design results in more accurate models; these subsequently lead to better controller performance. If the uncertainty in the estimates is small, the controller can be tuned more closely to its maximum performance since a smaller safety margin can be accommodated. Advanced control techniques such as model predictive control (Mayne 2014) which have gained increasing popularity in the industry rely heavily on model accuracy in maximising their capability. In model predictive control, the control signal is optimised at every time sample making use of the process model in predicting the output. Further to this, a carefully designed signal allows detailed analysis of linear contributions, nonlinear distortions and disturbances. This provides insights leading to well-informed decisions on whether to apply a linear controller to a dominantly linear process, a set of linear controllers to a linear parameter-varying process or a nonlinear control strategy to a process with significant nonlinear behaviour.

In light of time constraints in conducting identification tests in industry, the signal should be designed to allow a maximum amount of information to be collected within the available experimentation time. In a model predictive control project, the plant test and subsequent model identification often consume more than 50% of the total project time (Darby and Nikolaou 2012). In this aspect, broadband signals which simultaneously excite several frequencies within the bandwidth in a single experiment are advantageous compared to frequency-by-frequency testing. This is because for every experiment conducted, time is required for transient effects to decay and frequency-by-frequency testing requires a separate experiment for every frequency at which the system response is to be measured.

The objective of the identification should be taken into account as different objectives necessitate different signal designs. For instance, the identification of nonlinearities in a system requires a different signal than one used to identify the underlying linear dynamics of the same system with the effects of nonlinearities minimised. A trade-off is frequently needed in the light of conflicting requirements and practical constraints. A common scenario is in the selection of signal amplitude. By increasing the amplitude, the signal-to-noise ratio (SNR) at the output improves but doing so may excite unwanted nonlinearities. Another example is the selection of the number of signal levels. A larger number of levels lead to better frequency domain performance which is closer to the desired specifications. However, this typically results in lower power in the signal if the amplitude is kept unchanged. Thus, it is important to have many different signal designs to cater for different situations.

The focus of this book is on general purpose periodic broadband signals, where minimal knowledge of the system under test is assumed. These signals are widely applicable in many applications and specifically in identification for control. For pseudorandom signals, the classes considered belonging to PRB signals are maximum length binary (MLB), quadratic residue binary (QRB), Hall binary (HAB) and twin prime binary (TPB) signals; those belonging to ternary signals are quadratic residue ternary (QRT) signals, pseudorandom multilevel (PRML) signals having three levels, and direct synthesis ternary signals; and those belonging to multilevel signals are PRML signals generated from the appropriate Galois fields (GFs) which may have three or more levels. For computer-optimised signals, multisine, discrete interval binary (DIB), discrete interval ternary (DIT) and multilevel multiharmonic (MLMH) signals are suitable for general purpose testing. However, some designs which are optimised based on the availability of a larger amount of a priori information are considered in Sects. 3.5 and 4.3; these are most easily implemented using multisine signals. Optimal designs find applications in reidentification of existing control loops as well as in the identification of ill-conditioned systems. The advantages of using periodic signals will be explained in Sect. 1.6.

1.3 Specifications for the Identification of Linear Systems

1.3.1 Time Domain Specifications

Several common time domain specifications for perturbation signals are as follows:

- amplitude,
- number of signal levels,
- amplitude distribution, and
- signal period.

The maximum and minimum values of a perturbation signal are determined by the allowable range of the input transducer and that of the system output. The latter is influenced by safety considerations (for example, the temperature of a reactor), possible changes in the system at large amplitudes (for example, the output may enter a nonlinear range such as a saturation range in amplifier circuits), and maximum deviations from the nominal value (for example, in a production plant operating in closed loop).

For the identification of linear systems, a binary signal offers the best possible choice in terms of maximising signal power within amplitude constraints. However, this choice also comes with the least flexibility in terms of frequency domain specifications, and a trade-off may be necessary between the two conflicting requirements. Additionally, the number of signal levels may be limited by hardware constraints. Actuator limitation is a common cause; Barker and Godfrey (1999) described an application to a continuous hot-dip galvanising process for steel strip where a maximum of three levels could be applied. A similar point was made by Mohanty (2009), who applied ternary signals to excite a flotation column. Easier processing is also an important consideration. Pinter and Fernando (2010) utilised binary signals for the estimation of code division multiple access (CDMA) networks. Ternary signals were applied in optical CDMA networks allowing a simple encoder to be used in the control base station (Yang 2008). They were also applied for channel identification of fibre wireless uplinks in both single-user (Ng et al. 2011) and multi-user (Ng et al. 2016) scenarios. In Roinila et al. (2009), the use of a PRB signal allowed a simple circuitry to be implemented on a switched-mode converter. In Tan and Godfrey (2004), physical construction of the electronic nose system limited the signals to having at most four levels.

The amplitude distribution may be determined by the signal class in some cases, but in other cases, may be optimised by the user. For linear system identification, a binary or near-binary amplitude distribution will result in the highest SNR at the system output. The number of signal levels and amplitude distribution for the classes of signals considered in this book are summarised in Table 1.1.

The period of a perturbation signal depends on the sampling interval as well as the measurement time per period. In many cases, some indication of the dynamics of the system under study is available from preliminary step tests or historical records.

Table 1.1 Number of signal levels and amplitude distribution for various classes of signals

Class	Number of levels	Amplitude distribution
MLB, QRB, HAB, TPB	2	Almost uniformly distributed across the two levels
Inverse-repeat MLB, QRB, HAB, TPB	2	Uniformly distributed across the two levels
QRT and inverse-repeat QRT	3	Near-binary, with very small number of occurrences of the level 0, and equal number of occurrences of the other two levels
PRML and truncated PRML	Depends on GF, with some user flexibility	Depends on harmonic specification
Direct synthesis ternary	3	Uniformly distributed across the three levels
Suboptimal direct synthesis ternary	3	Almost uniformly distributed across the three levels
Multisine	Practically infinite	Gaussian for random-phase design, increasing resemblance to binary signals with increasing crest factor minimisation
DIB	2	Almost uniformly distributed across the two levels
DIT	3	Depends on harmonic specification
MLMH	Small number of levels	Depends on harmonic specification
Gallev	Small number of levels	Depends on harmonic specification

Several guidelines exist on the selection of the sampling interval and measurement period. A simple choice is to select the sampling interval t_s such that

$$t_s \leq \frac{\text{minimum time constant}}{5} \tag{1.2}$$

and the measurement period T_N such that

$$T_N \geq 5 \times \text{maximum time constant}. \tag{1.3}$$

The signal period N is then computed from

$$N = T_N / t_s. \tag{1.4}$$

The choice of sampling interval sets the sampling frequency f_s since

$$f_s = 1/t_s \tag{1.5}$$

whereas the choice of measurement period sets the frequency resolution $f_{resolution}$ since

$$f_{resolution} = 1/T_N. \tag{1.6}$$

However, the effective frequency resolution may be larger if some harmonics are suppressed.

For broadband signals with uniform discrete Fourier transform (DFT) magnitudes such as PRB signals, it is recommended to set the sampling frequency equal to 2.5 times the maximum frequency of interest (Pintelon and Schoukens 2012). Note that throughout this book, it is normally assumed that the sampling interval is the same as the clock pulse interval. In any case where this is not so, it will be explicitly mentioned.

1.3.2 Frequency Domain Specifications

Several common frequency domain specifications for perturbation signals are as follows:

- bandwidth,
- harmonic spacing, such as linear or (quasi-)logarithmic,
- amplitude spectrum or DFT magnitude of the excited harmonics, and
- with or without zero-order hold (ZOH) pre-compensation.

Computer-optimised signals offer greater flexibility in terms of frequency domain specifications compared with pseudorandom signals. Pseudorandom signals have uniform DFT magnitude across the whole spectrum with linearly spaced excited harmonics which may also incorporate harmonic suppression. (Note that it is possible, via an incorrect design, to obtain non-uniform spectrum, but this is not really what these signals are intended for.) Computer-optimised signals such as multisine, DIB, DIT and MLMH signals are able to accommodate low-pass and bandpass designs as well as linearly spaced and (quasi-)logarithmically spaced excited harmonics. Multisine signals can exactly meet any specification in terms of amplitude spectrum. In particular, the amplitude spectrum can be shaped to match that of an optimal design (see Sect. 3.5). DIB, DIT and MLMH signals are not able to exactly satisfy specifications on amplitude spectrum due to their limited number of levels.

It is interesting to note that the ideal amplitude spectrum depends on whether a parametric model or a nonparametric model is to be estimated. In the former case, more power should be inserted at frequencies which contribute most to the system parameters. This is typically where the system gain is large as the measurements at these frequencies have smaller uncertainties and will result in more accurate para-

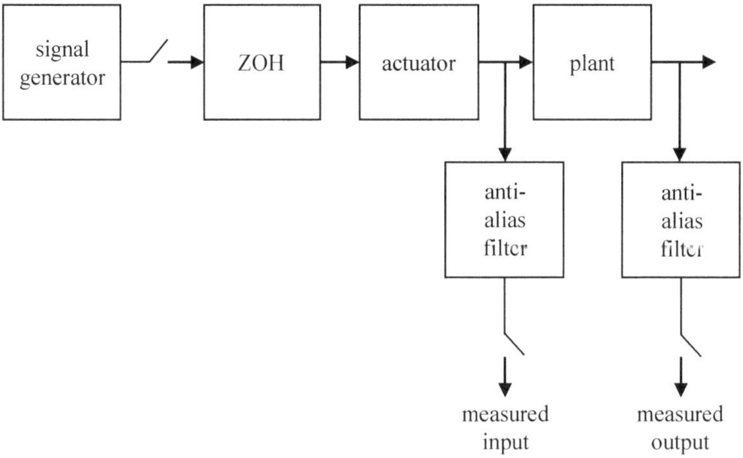

Fig. 1.1 Band-limited identification setting

metric estimates. In the latter case, power should be distributed so that a minimum predefined accuracy can be obtained across the bandwidth of interest.

Computer-optimised signals such as multisine, DIB, DIT and MLMH signals may be designed either with or without ZOH pre-compensation for the amplitude spectrum. For continuous-time signals, the choice depends on the intersampling behaviour. There are two popular assumptions (Pintelon and Schoukens 2012; Schoukens et al. 2018). The first is band-limited assumption, where the band-limited data have no power above the Nyquist frequency due to the presence of anti-alias filters. The signal from the generator is sampled with a very high sampling frequency so that the ZOH has little effect on the spectrum. As such, no ZOH pre compensation is required in the signal design. The identification setting is shown in Fig. 1.1 where it can be seen that both the input and output signals are measured after the anti-alias filters. Additionally, in a practical system, noise may enter at various locations in Fig. 1.1.

The second assumption is ZOH assumption, where the signal is piecewise constant between samples. This is the typical assumption when performing identification for discrete controller design. The identification setting is shown in Fig. 1.2, where in a practical system, noise may also enter at various locations. There will be high-frequency components due to the ZOH assumption and no anti-alias filter is used. Since the input signal at the generator is known, the identification is done for the whole system which includes the ZOH, actuator and plant. For this setting, the input signal may be specified to include ZOH pre-compensation, in order to compensate for the $(\sin^2 x)/x^2$ power spectrum envelope of the ZOH. With the ZOH pre-compensation, the required amplitude spectrum appears at the input to the actuator.

In the case of a discrete-time signal being required, intersampling behaviour is not applicable and the signal should be designed without ZOH pre-compensation.

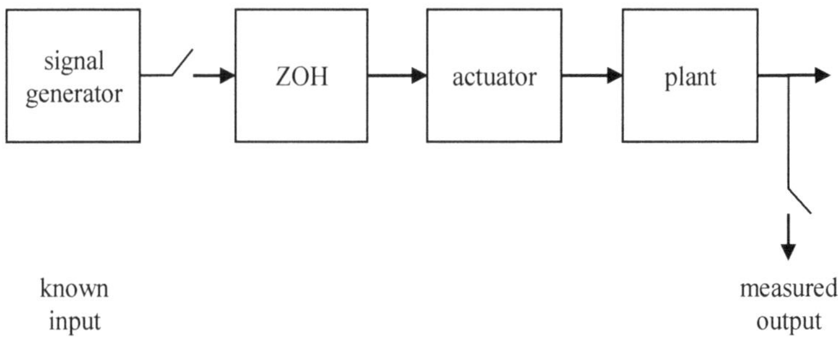

Fig. 1.2 ZOH identification setting

The focus of this book is on the design of continuous perturbation signals in the ZOH identification setting using the assumption that the signal is piecewise constant between samples, due to its relevance to process control. However, the design of continuous perturbation signals with the band-limited assumption and the design of discrete perturbation signals are also covered, to a lesser extent.

1.4 Performance Measures for the Identification of Linear Systems

For the identification of the underlying linear dynamics of a system, the perturbation signal should satisfy two main criteria. The first is that the signal amplitude should be sufficiently small in order to minimise the effects of nonlinearities. The second is that the power contained in the specified harmonics should be sufficiently large in order to achieve an acceptable SNR. Since these are conflicting requirements, a prime purpose in signal design is to maximise the power in the specified harmonics within the amplitude constraints. For comparing different signals against this objective, it is essential to have a suitable performance measure; this is discussed in Sect. 1.4.1. Note that for the identification of nonlinear systems, modifications to the performance measure may be needed depending on the application. Additional requirements such as persistency of excitation in both amplitude and frequency domains may be required. A frequency domain measure is considered in Sect. 1.4.2.

1.4.1 Performance Index for Perturbation Signals

The dispersion of a stationary signal u (Godfrey et al. 1999) can be expressed by a normalised measure λ_u defined as

$$\lambda_u = \frac{u_{\max} - u_{\min}}{2\sigma_u} = \frac{u_{\max} - u_{\min}}{2\sqrt{u_{\mathrm{rms}}^2 - u_{\mathrm{mean}}^2}}. \tag{1.7}$$

The signal may be continuous, with $u = u(t)$ or discrete, with $u = u(i)$. In Eq. 1.7, $u_{\max}$, $u_{\min}$, σ_u, u_{mean} and u_{rms} denote the maximum, minimum, standard deviation, mean and root mean square (RMS) of the signal u, respectively. In particular, u_{mean} and u_{rms} are defined by

$$u_{\mathrm{mean}} = \mathrm{E}[u(t)] \text{ or } \mathrm{E}[u(i)] \tag{1.8}$$

and

$$u_{\mathrm{rms}} = \sqrt{\mathrm{E}[u^2(t)]} \text{ or } \sqrt{\mathrm{E}[u^2(i)]}. \tag{1.9}$$

The dispersion λ_u has a lower limit of unity but no upper limit. A signal with the least possible dispersion is a binary signal with a uniform amplitude distribution between the levels $u_{\max}$ and $u_{\min}$. λ_u is independent of both the mean and the amplitude scale of u, which makes it an advantage over other measures such as crest factor, peak factor and form factor, defined by

$$\text{crest factor} = u_{\mathrm{peak}}/u_{\mathrm{rms}}, \tag{1.10}$$

with $u_{\mathrm{peak}} = \max[|u|]$,

$$\text{peak factor} = \frac{u_{\max} - u_{\min}}{2\sqrt{2}u_{\mathrm{rms}}}, \tag{1.11}$$

and

$$\text{form factor} = u_{\mathrm{rms}}/\mathrm{E}[|u|]. \tag{1.12}$$

These measures were originally used to quantify sine wave distortions, and were not designed with perturbation signals specifically in mind. They have the additional drawback of being defined differently by different authors (Godfrey et al. 1999). Crest factor, peak factor and λ_u also share the common disadvantage that they have finite lower limits (for signals with the least possible dispersion), but no upper limits. However, crest factor is a widely used measure.

It is preferable to use performance indices that range from 0% (worst possible performance) to 100% (best possible performance, corresponding to least dispersion). A suitable measure, the performance index for perturbation signals (PIPS), is inversely proportional to λ_u and is given by (Godfrey et al. 1999)

$$\mathrm{PIPS} = 100\lambda_u^{-1} = \frac{200\sqrt{u_{\mathrm{rms}}^2 - u_{\mathrm{mean}}^2}}{u_{\max} - u_{\min}}\,\%\,,\; u_{\max} > u_{\min}. \tag{1.13}$$

PIPS is independent of the signal mean and amplitude scale. It is 100% for a signal with the best possible performance. A PIPS of 100% can be achieved by binary signals for which u_{max} and u_{min} have equal duration or number of occurrences (uniform amplitude distribution).

PIPS is applicable to all stationary signals, but the special case of periodic signals is particularly important. If u is a periodic signal, with period $N \times t_s$ when u is continuous or period N when u is discrete, then the expectations in Eqs. 1.8 and 1.9 are obtained by averaging over a period, so that

$$u_{\mathrm{mean}} = \frac{1}{Nt_s}\int_0^{Nt_s} u(t)\mathrm{d}t = \frac{1}{N}\sum_{i=1}^{N} u(i) \tag{1.14}$$

and

$$u_{\mathrm{rms}} = \sqrt{\frac{1}{Nt_s}\int_0^{Nt_s} u^2(t)\mathrm{d}t} = \sqrt{\frac{1}{N}\sum_{i=1}^{N} u^2(i)}. \tag{1.15}$$

In this case, PIPS can also be expressed in terms of the frequency content of u. If the DFT of the discrete signal $u(i)$ is $U(k)$, defined as

$$U(k) = \sum_{i=1}^{N} u(i)\exp\left(-\frac{\mathrm{j}2\pi ki}{N}\right) \tag{1.16}$$

then using Parseval's theorem

$$\sum_{i=1}^{N} u^2(i) = \frac{1}{N}\sum_{k=0}^{N-1} |U(k)|^2 \tag{1.17}$$

leads to

$$u_{\mathrm{rms}}^2 = \frac{1}{N}\sum_{i=1}^{N} u^2(i) = \frac{1}{N^2}\sum_{k=0}^{N-1} |U(k)|^2 = \frac{1}{N^2}\sum_{k=1}^{N-1} |U(k)|^2 + u_{\mathrm{mean}}^2. \tag{1.18}$$

Substituting Eq. 1.18 into Eq. 1.13 results in

$$\mathrm{PIPS} = \frac{200\sqrt{\sum_{k=1}^{N-1} |U(k)|^2}}{N(u_{\max} - u_{\min})}\%. \tag{1.19}$$

The absence of the zero harmonic from Eq. 1.19 confirms that PIPS is independent of the mean of u.

PIPS is an application-independent quality measure which may, however, be developed further to suit specific applications. Consider for example the identification of a continuous system where the intersampling behaviour follows the ZOH assumption. Under this setting, modulation by the frequency response of the ZOH through which the continuous signal $u(t)$ is generated from the discrete signal $u(i)$ reduces the effectiveness of the harmonics due to the $(\sin^2 x)/x^2$ power spectrum envelope of the ZOH. In this case, PIPS (effective), or simply PIPSE, may be derived for a periodic perturbation signal u used for the identification of a continuous system with the output sampled at the same rate as the input (Godfrey et al. 1999).

In a bilateral spectrum that includes both positive and negative frequencies, the power of the kth harmonic of $u(t)$ is given by

$$|C_u(k)|^2 = \left| \frac{\sin(\pi k/N)}{(\pi k/N)} \frac{U(k)}{N} \right|^2. \tag{1.20}$$

Thus, the power in the kth harmonic in a unilateral spectrum (which includes non-negative frequencies only) is given by

$$|C_u'(k)|^2 = \begin{cases} |C_u(k)|^2 = \left|\frac{U(k)}{N}\right|^2 & k = 0 \\ 2|C_u(k)|^2 = 2\left|\frac{\sin(\pi k/N)}{(\pi k/N)} \frac{U(k)}{N}\right|^2 & k > 0 \end{cases}. \tag{1.21}$$

If the first R nonzero harmonics are specified for the identification, then the appropriate PIPSE is given by (Godfrey et al. 2005)

$$\text{PIPSE} = \frac{200\sqrt{\sum_{k=1}^{R} |C_u'(k)|^2}}{(u_{\max} - u_{\min})} \%, \; R < N/2. \tag{1.22}$$

PIPSE is maximised when the specified harmonics contain the greatest proportion of the signal power. This occurs if R is $(N-2)/2$ when N is even and $(N-1)/2$ when N is odd.

In applications where the perturbation signal incorporates harmonic suppression, PIPSE may be modified accordingly. For example, if only odd harmonics are specified, then $|C_u'(k)|^2$ may be replaced by $|C_u'(2k-1)|^2$ in Eq. 1.22.

1.4.2 Frequency Domain Measure of Signal Quality

The use of PIPSE alone as a measure of signal quality does not take into account the possibility that the power contained in one of the specified harmonics is low; this could result in low accuracy of estimation of the frequency response at that frequency (Godfrey et al. 1999). To reflect this possibility, a frequency domain measure is needed. This measure is the minimum ratio $E_{\min}$ at any of the specified harmonics between the actual harmonic amplitude and the specified harmonic amplitude defined by

$$E_{\min} = \min_{\text{specified } k} \left(\frac{|C'_u(k)|}{|C'_u(k)|_{\text{specified}}} \right). \tag{1.23}$$

The measure $E_{\min}$ is well suited for optimising signals of a given type. However, for comparing signals of different types, modification is required. For computer-optimised signals, $|C'_u(k)|_{\text{specified}}$ is an input to the optimisation algorithm and is not a signal parameter. As a consequence, an algorithm could always be arranged to ensure that $\left|C'_u(k)\right|$ is larger than $|C'_u(k)|_{\text{specified}}$, so that $E_{\min}$ would never be less than 1, hence negating the principle on which it is based. A simple refinement circumvents the problem without changing the principle. As the average power in the specified harmonics is ideally $|C'_u(k)|^2_{\text{specified}}$, for the same signal and application for which PIPSE is defined in Eq. 1.22,

$$|C'_u(k)|^2_{\text{specified}} = \frac{1}{R} \sum_{k=1}^{R} |C'_u(k)|^2, \; R < N/2. \tag{1.24}$$

With this refinement, an $E_{\min}$ (effective) which is the effective minimum ratio between the actual amplitude and the specified amplitude at any of the specified harmonics, or simply EMINE, can be defined as (Godfrey et al. 2005)

$$\text{EMINE} = 100 \min_{k=1,2,\ldots,R} \frac{\left|C'_u(k)\right|}{\sqrt{\frac{1}{R} \sum_{k=1}^{R} \left|C'_u(k)\right|^2}} \%, \; R < N/2. \tag{1.25}$$

EMINE is independent of the signal mean and amplitude scale. It has a lower limit of 0% for a signal with the worst possible performance and an upper limit of 100% for a signal with the best possible performance. In the case where the signal incorporates harmonic suppression, Eq. 1.25 is applied only at the specified harmonics.

Since the power in the specified harmonics appears in the numerator of PIPSE and in the denominator of EMINE, signal design using these quality measures will inevitably be a compromise. In general, PIPSE is considered to be the more important of the two, because EMINE reflects a lack of power in only one of the specified harmonics but it does not provide indication of power at the rest of the specified harmonics. It is worth highlighting that both PIPSE and EMINE are defined only for the ZOH identification setting.

1.5 Harmonic Suppression in the Presence of Nonlinear Distortions

Harmonic suppression refers to the suppression of some harmonics, by setting the power at those harmonics to zero. There are two main reasons for doing so. The first is to allow detection of the effects of nonlinear distortions in the output at the detection lines, which are the harmonics corresponding to the suppressed harmonics at the input. The second is to allow the effects of nonlinear distortions to be filtered out either completely or partially, so as to eliminate or reduce their effects on the linear estimate.

The most common form of harmonic suppression is suppressing harmonic multiples of two such that

$$U(k) = 0 \forall k \in P, \tag{1.26}$$

where $P = \{2m | 0 \leq m < N/2, m \in Z\}$. This enables the effects of even-order nonlinearities on the linear estimate to be completely eliminated (Pintelon and Schoukens 2012). Additionally, harmonic multiples of three may be suppressed so that the effects of odd-order nonlinearities on the linear estimate are reduced (Tan et al. 2015). Note that it is theoretically not possible to completely eliminate the effects of odd-order nonlinearities on the linear estimate. The specification with harmonic multiples of two and three suppressed is defined by

$$U(k) = 0 \forall k \in P \cup Q, \tag{1.27}$$

where $Q = \{3m | 0 \leq m < N/3, m \in Z\}$.

The nth power of the sum of F number of sinusoids is given by (Schoukens et al. 2012)

$$\left(\sum_{k=1}^{F} A_k \cos(k\omega_0 t + \phi_k) \right)^n = \sum_{k_1, k_2, \ldots, k_n = 1}^{F} \prod_{i=1}^{n} A_{k_i} \cos(k_i \omega_0 t + \phi_{k_i}). \tag{1.28}$$

To illustrate this with some examples, assume for simplicity that the phases of the sinusoids are zero. For a signal with two harmonic components at ω_0 and $2\omega_0$ (harmonics 1 and 2), its second power is

$$\begin{aligned} &(A_1 \cos \omega_0 t + A_2 \cos 2\omega_0 t)^2 \\ &= \frac{A_1^2}{2}(1 + \cos 2\omega_0 t) + A_1 A_2 (\cos \omega_0 t + \cos 3\omega_0 t) + \frac{A_2^2}{2}(1 + \cos 4\omega_0 t) \end{aligned} \tag{1.29}$$

and harmonics 0, 1, 2, 3 and 4 appear in the output. The third power of the signal can be determined in a similar way to be

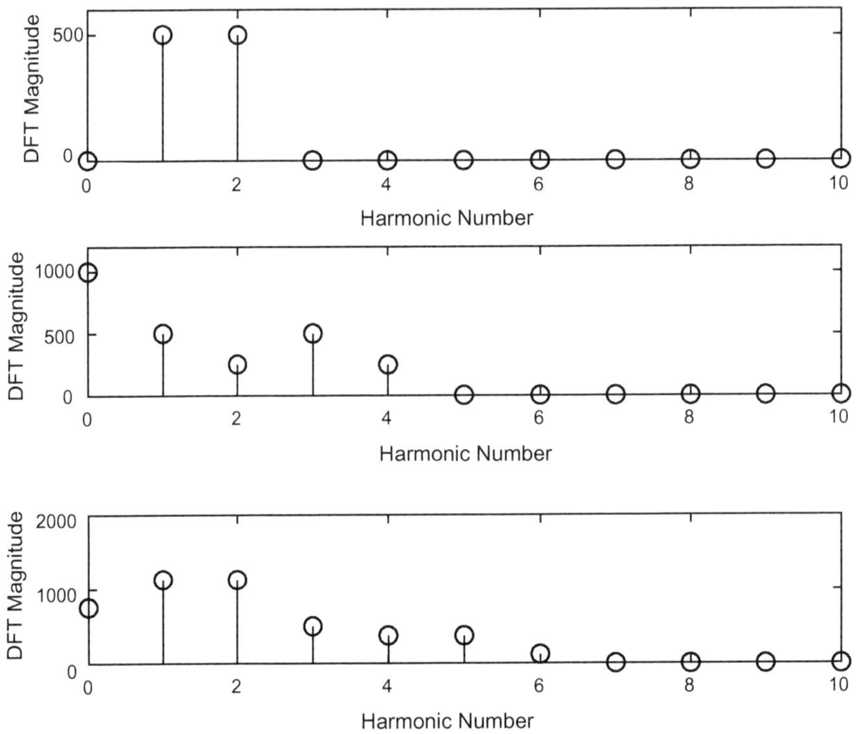

Fig. 1.3 Top: DFT of signal u with harmonics 1 and 2; middle: DFT of u^2; bottom: DFT of u^3

$$
\begin{aligned}
&(A_1 \cos\omega_0 t + A_2 \cos 2\omega_0 t)^3 \\
&\quad = \frac{A_1^3}{4}(3\cos\omega_0 t + \cos 3\omega_0 t) + \frac{A_1^2 A_2}{4}(3 + 6\cos 2\omega_0 t + 3\cos 4\omega_0 t) \\
&\quad + \frac{A_1 A_2^2}{4}(6\cos\omega_0 t + 3\cos 3\omega_0 t + 3\cos 5\omega_0 t) \\
&\quad + \frac{A_2^3}{4}(3\cos 2\omega_0 t + \cos 6\omega_0 t),
\end{aligned} \tag{1.30}
$$

where harmonics 0, 1, 2, 3, 4, 5 and 6 appear in the output. This is shown in Fig. 1.3 for $A_1 = A_2 = 1$. (The scaling on the vertical axis depends on the number of points taken for the DFT.) Note that the highest harmonic at the output is given by the order of the nonlinearity multiplied by the highest harmonic in the input.

Consider now the case if the input has harmonics 1 and 3 instead. The second power is given by

$$
\begin{aligned}
&(A_1 \cos\omega_0 t + A_3 \cos 3\omega_0 t)^2 \\
&= \frac{A_1^2}{2}(1 + \cos 2\omega_0 t) + A_1 A_3(\cos 2\omega_0 t + \cos 4\omega_0 t) + \frac{A_3^2}{2}(1 + \cos 6\omega_0 t)
\end{aligned} \tag{1.31}
$$

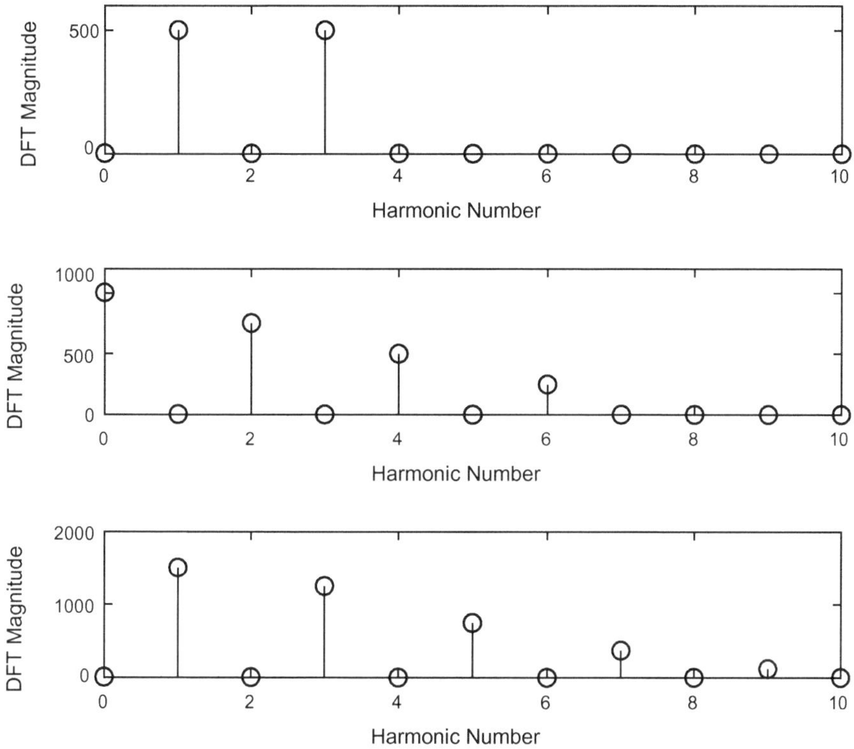

Fig. 1.4 Top: DFT of signal u with harmonics 1 and 3; middle: DFT of u^2; bottom: DFT of u^3

and the output contains harmonics 0, 2, 4 and 6 which are all even harmonics. Similarly,

$$\begin{aligned}(A_1 \cos \omega_0 t + A_3 \cos 3\omega_0 t)^3 \\ = \frac{A_1^3}{4}(3 \cos \omega_0 t + \cos 3\omega_0 t) + \frac{A_1^2 A_3}{4}(3 \cos \omega_0 t + 6 \cos 3\omega_0 t + 3 \cos 5\omega_0 t) \\ + \frac{A_1 A_3^2}{4}(6 \cos \omega_0 t + 3 \cos 5\omega_0 t + 3 \cos 7\omega_0 t) + \frac{A_3^3}{4}(3 \cos 3\omega_0 t + \cos 9\omega_0 t)\end{aligned} \quad (1.32)$$

and the output contains harmonics 1, 3, 5, 7 and 9 which are all odd harmonics. This is depicted in Fig. 1.4 for $A_1 = A_3 = 1$.

Noting that even-order combinations of odd numbers result in even numbers and odd-order combinations of odd numbers result in odd numbers, the effects of odd- and even-order nonlinearities can be separated if the input has harmonic multiples of two suppressed. More detailed frequency domain analysis can be performed with further harmonic suppression, as shown in Sect. 5.4. It should also be noted that the

Table 1.2 Harmonic suppression for various classes of signals

Class	Harmonic suppression
MLB, QRB, HAB, TPB, QRT	None
Inverse-repeat MLB, QRB, HAB, TPB, QRT	Multiples of two
PRML and truncated PRML	Depends on GF, harmonic multiples of k may be suppressed if the signal period is an integer multiple of k
Direct synthesis ternary and suboptimal direct synthesis ternary	Multiples of two and three
Multisine	Harmonic multiples of k may be suppressed if the signal period is an integer multiple of k
DIB	Multiples of two
DIT	Multiples of two, and multiples of two and three
MLMH	Depends on the number of signal levels
Gallev	Depends on GF, harmonic multiples of k may be suppressed if the signal period is an integer multiple of k

analysis is much more involved if the system is identified in closed loop as the input to the process now contains other harmonic components due to the feedback. The interested reader is referred to Chap. 6 of Schoukens et al. (2012) for further details.

The types of harmonic suppression for the classes of signals considered in this book are summarised in Table 1.2.

1.6 General Comparison Between Periodic and Non-periodic Signals

Non-periodic (aperiodic) perturbation signals are frequently used for identification despite periodic signals offering many advantages over non-periodic ones. Three common non-periodic signals, namely impulse signals, step signals and random noise, will be explained followed by a comparison between periodic and non-periodic signals.

1.6.1 Impulse Signals

Impulse signals are commonly utilised as excitation signals since they are easy to apply and they allow the impulse response of the system under test to be measured directly. They have a further advantage that they are the shortest possible form of excitation signal. In practice, they are pulse signals with very short pulse durations. Traditionally, in the area of mechanical engineering where the use of such inputs is quite popular, the impulse is inflicted by hitting the specimen with a hammer. The input spectrum is white. The main drawback is that it may be difficult to apply an impulse of sufficiently large amplitude in order to accurately measure the impulse response in the presence of noise.

Impulse excitation has recently been applied to measure the fundamental resonant frequency of composite laminates for the computation of Young's modulus (Paolino et al. 2017), to identify room acoustics (Murphy et al. 2014), to measure the impulse response of a grounding grid so as to calculate its impedance (Visacro et al. 2013) and to study the optical excitation of matter with linearly polarised femtosecond pulses (Makino et al. 2016). Berezvai et al. (2018) investigated the use of airsoft pellets and bearing balls fired using a special pneumatic gun as means of providing impulse excitations in place of hammers.

1.6.2 Step Signals

Step inputs are frequently used due to them being simple and practical to apply. No signal generator is needed to perform a step test (Liu and Gao 2012). They are indeed very useful for obtaining basic information about the system, for example, the gain and settling time. Such information can be used to subsequently design a broadband signal for detailed characterisation. Step inputs can aid detection of certain types of nonlinearities, such as those in a Hammerstein structure (where a static nonlinearity precedes linear dynamics) and those in a Wiener structure (where a static nonlinearity follows linear dynamics), bilinear behaviour and direction-dependent behaviour. The detection of direction-dependent behaviour of an electronic nose system through step tests is described in Sect. 6.2.1. Step tests were also applied by Chook and Tan (2007) for the identification of a bilinear electric resistance furnace.

Step tests have two main disadvantages. First, long experimentation time is needed if several step tests of different amplitudes are required which is typically the case when nonlinearities are to be characterised. Second, the signal has very little high-frequency content, as can be seen from Fig. 1.5, where the DFT is taken across 100 points.

1.6.3 Random Noise

Random white noise can be generated for various amplitude distributions such as Gaussian distribution and uniform distribution. It can also be imposed to have only two values, by passing non-binary noise through a relay. A binary noise provides higher PIPS compared with noise with a larger number of levels.

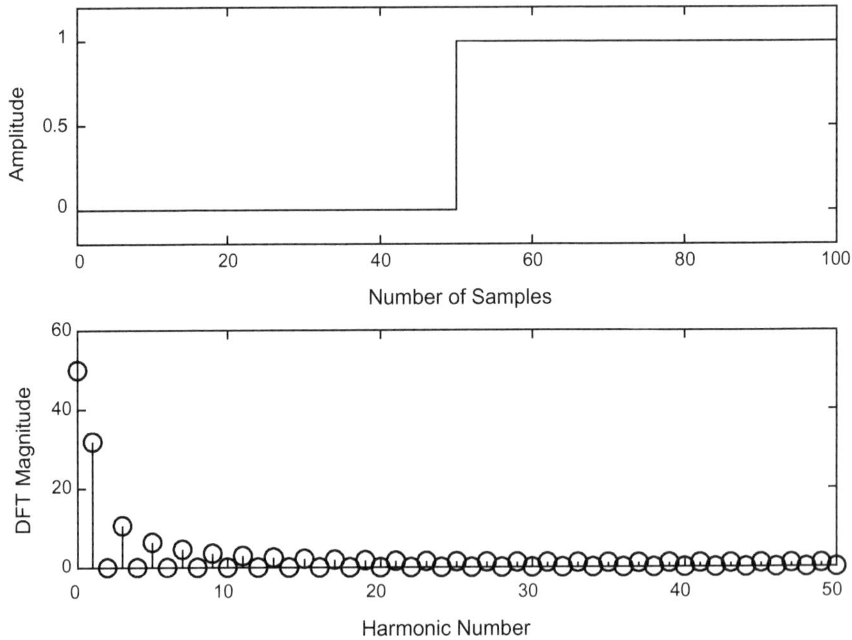

Fig. 1.5 Step input

An important advantage of random noise is its ease of generation. However, a significant drawback is the non-uniformity of the amplitude spectrum. The power at some harmonics is very small (small EMINE) leading to poor estimation at the corresponding frequencies. An example is shown in Fig. 1.6 for one particular realisation of a Gaussian distributed random noise with RMS value of 1. The corresponding binary noise generated from the Gaussian noise is shown in Fig. 1.7, scaled to RMS value of 1. Different realisations will result in different amplitude spectrum and hence, many averages are necessary to smooth the spectrum. Another shortcoming is leakage problems in the frequency domain. The effects of leakage depend on the window used.

The spectrum can be shaped to some extent by passing the noise through a digital filter. However, the output of the filter will have an amplitude distribution which tends to Gaussian even if the input to the filter is non-Gaussian (Schoukens et al. 2012). The noise signals of Figs. 1.6 and 1.7, after being passed through a Butterworth low-pass filter (and scaling to achieve RMS value of 1), result in the noise signals of Figs. 1.8 and 1.9. It can be observed from Fig. 1.9 that the noise signal is no longer binary. In fact, it resembles the Gaussian noise of Fig. 1.8.

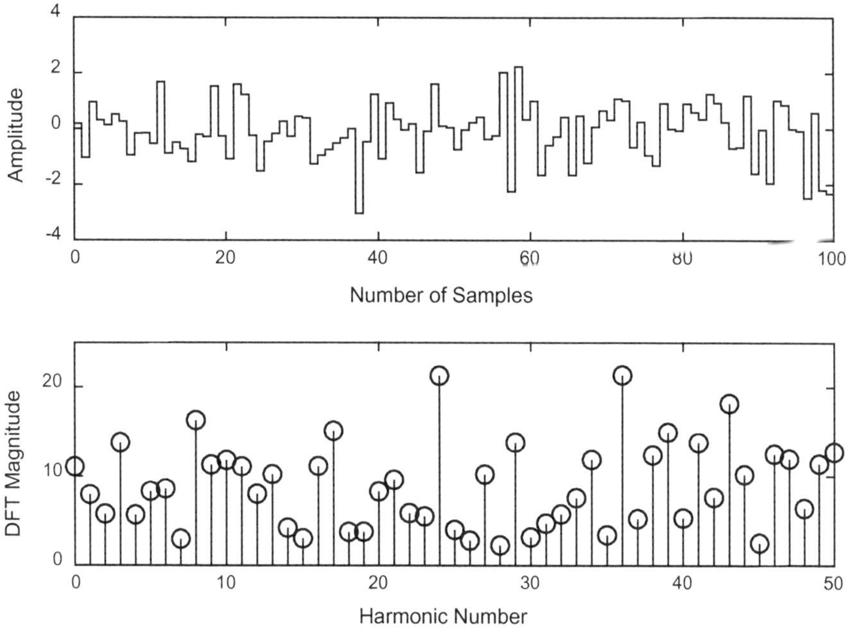

Fig. 1.6 Random Gaussian white noise

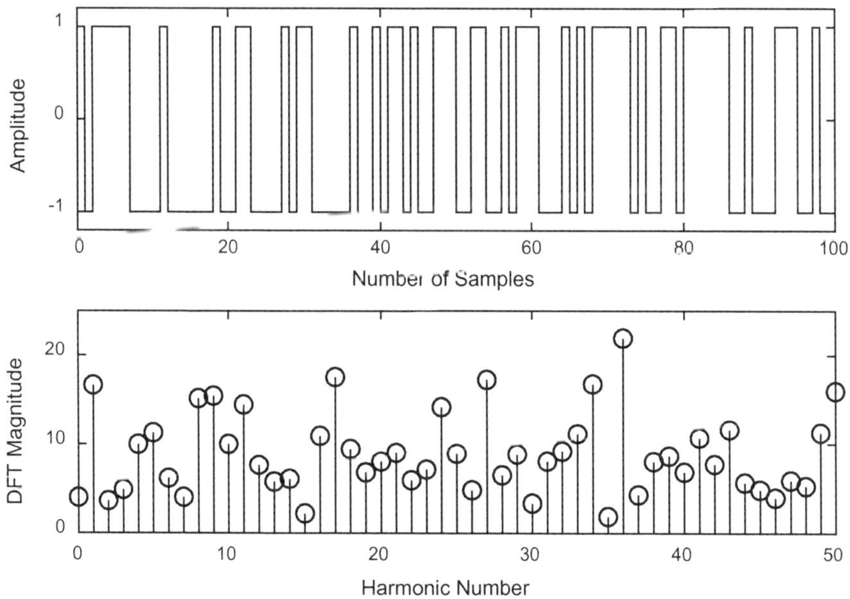

Fig. 1.7 Random binary white noise

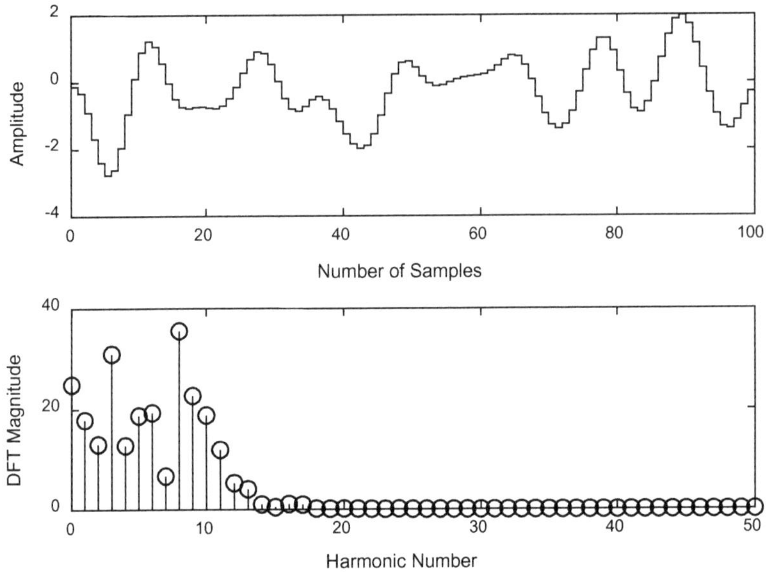

Fig. 1.8 Result of passing random Gaussian white noise through a low-pass filter

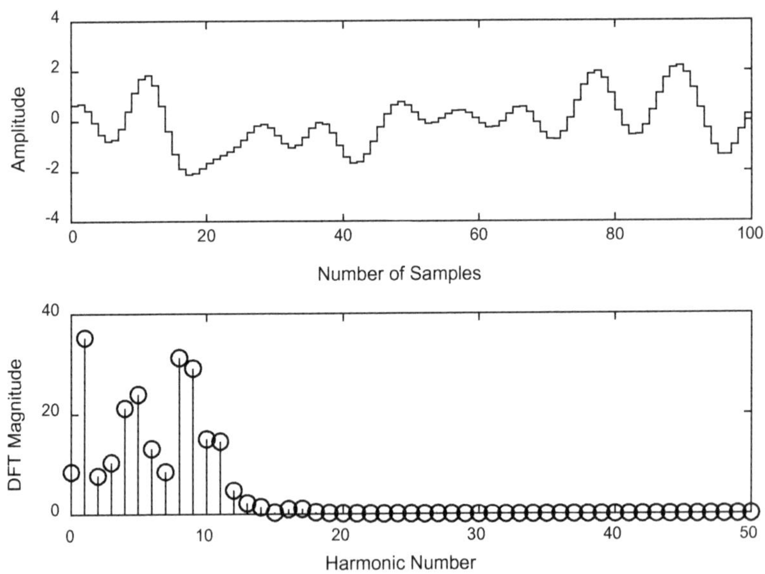

Fig. 1.9 Result of passing random binary white noise through a low-pass filter

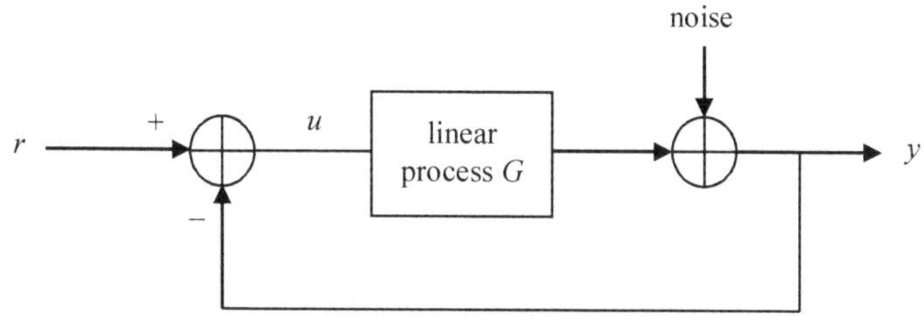

Fig. 1.10 Process in closed loop

1.6.4 Comparison Between Periodic and Non-periodic Signals

A comparison is given in Table 1.3. In general, periodic signals offer many advantages compared with non-periodic ones in terms of avoidance of leakage effects, having a deterministic spectrum, straightforward averaging, incorporation of harmonic suppression, applicability of the direct method for closed-loop identification and simple handling of multirate sampling.

Of particular importance is the fact that for closed-loop identification of a process (see Fig. 1.10), the direct method can be applied if multiple periods of a periodic signal are utilised. In the direct method, the process frequency response function (FRF) G is measured directly from u to y in the same way as in the open loop case, by taking the ratio of the cross-power spectrum between y and u, denoted by S_{uy}, to the auto-power spectrum of u, denoted by S_{uu}. The FRF at harmonic k is given by

$$G(k) = \frac{S_{uy}(k)}{S_{uu}(k)} = \frac{\mathrm{E}[Y(k)\overline{U}(k)]}{\mathrm{E}[|U(k)|^2]}, \tag{1.33}$$

where $\overline{U}$ represents the complex conjugate of U. The indirect method needs to be utilised if non-periodic signals are applied. Otherwise, a bias will appear due to the feedback of the noise causing a correlation between the input to the process and the noise. In the indirect method, two separate FRFs are estimated, the first from r to u giving G_{ru} and the second from r to y giving G_{ry}. The FRF G is then obtained from

$$G = \frac{G_{ry}}{G_{ru}} = \frac{\left(\frac{S_{ry}(k)}{S_{rr}(k)}\right)}{\left(\frac{S_{ru}(k)}{S_{rr}(k)}\right)} = \frac{S_{ry}(k)}{S_{ru}(k)} = \frac{\mathrm{E}[Y(k)\overline{R}(k)]}{\mathrm{E}[U(k)\overline{R}(k)]}. \tag{1.34}$$

Multirate sampling refers to a situation where different signals in a system are being sampled at different rates. According to Xue et al. (2018), it is very common in sampled-data systems. For example, the output may have a lower sampling rate compared to the input. Periodic signals allow simple handling of multirate sampling

Table 1.3 Comparison between periodic and non-periodic signals

Characteristics	Periodic signals	Non-periodic signals
Leakage in the DFT	No leakage if integer periods of the signal are measured	Leakage appears for random noise. For impulse and step inputs, there is no leakage if the measurement time is sufficiently long such that the output signal reaches steady-state
Variation in amplitude spectrum	No variation and the signal is deterministic once it has been designed	Variation between different realisations
Averaging to improve SNR	Simple averaging can be done if more than one period is measured. The effects of noise reduce inversely proportional to $\sqrt{M}$ (the standard deviation of the FRF estimate reduces by a factor of $\sqrt{M}$), where M is the number of periods measured	Averaging can be done to smooth the spectrum over successive realisations at the cost of reduced frequency resolution
Harmonic suppression	Harmonic suppression can be easily incorporated in the signal and this is very useful for characterising contributions of different components in the system	Not possible
Identification in closed loop	Direct method of estimating the process FRF can be applied if multiple periods are measured	Direct method of estimating the process FRF cannot be applied due to the feedback of process noise being correlated with the input hence causing a bias (except for some special cases). Indirect method should be used instead
Multirate sampling	Simple handling through multiple experiments with the input having different shifts in each experiment	Methods such as polynomial transformation may be applied (Lu and Fisher 1989)

by conducting multiple experiments with the input having different shifts in each experiment (Tan 2018).

In light of the many advantages of periodic signals over non-periodic signals, the former should be the prime choice in most applications. However, non-periodic signals remain attractive choices particularly for preliminary tests due to them being easy to generate and apply.

References

Barker HA, Godfrey KR (1999) System identification with multi-level periodic perturbation signals. Control Eng Pract 7:717–726

Berezvai S, Kossa A, Bachrathy D, Stepan G (2018) Numerical and experimental investigation of the applicability of pellet impacts for impulse excitation. Int J Impact Eng 115:19–31

Chang JA, Owen AG, Zaman M, Griffin AWJ (1968) Dynamic modelling of a four stand cold rolling steel mill using multilevel pseudo-random sequences. Meas Control 1:T80–T84

Chook KC, Tan AH (2007) Identification of an electric resistance furnace, IEEE Trans Instrum Meas 56:2262–2270

Darby ML, Nikolaou M (2012) MPC: Current practice and challenges. Control Eng Pract 20:328–342

Godfrey KR (1969a) The theory of the correlation method of dynamic analysis and its application to industrial processes and nuclear power plant. Meas Control 2:T65–T72

Godfrey KR (1969b) Dynamic analysis of an oil-refinery unit under normal operating conditions. Proc Inst Electr Eng 116:879–888

Godfrey KR, Barker HA, Tucker AJ (1999) Comparison of perturbation signals for linear system identification in the frequency domain. IEE Proc Control Theory Appl 146:535–548

Godfrey KR, Tan AH, Barker HA, Chong B (2005) A survey of readily accessible perturbation signals for system identification in the frequency domain. Control Eng Pract 13:1391–1402

Liu T, Gao F (2012) Industrial process identification and control design: step-test and relay-experiment-based methods. Springer, London, UK

Lu W, Fisher DG (1989) Least-squares output estimation with multirate sampling. IEEE Trans Autom Control 34:669–672

Makino K, Saito Y, Fons P, Kolobov AV, Nakano T, Tominaga J, Hase M (2016) Anisotropic lattice response induced by a linearly-polarized femtosecond optical pulse excitation in interfacial phase change memory material. Sci Rep 6: Article 19758

Mayne DQ (2014) Model predictive control: recent developments and future promise. Automatica 50:2967–2986

Mohanty S (2009) Artificial neural network based system identification and model predictive control of a flotation column. J Process Control 19:991–999

Murphy DT, Southern A, Savioja L (2014) Source excitation strategies for obtaining impulse responses in finite difference time domain room acoustics simulation. Appl Acoust 82:6–14

Ng YH, Tan AH, Chuah TC (2011) Channel identification of concatenated fiber-wireless uplink using ternary signals. IEEE Trans Veh Technol 60:3207–3217

Ng YH, Tan AH, Chuah TC (2016) Iterative channel estimation for multiuser fibre-wireless uplink exploiting semi-correlated ternary signals. IET Signal Proc 10:395–403

Paolino DS, Geng H, Scattina A, Tridello A, Cavatorta MP, Belingardi G (2017) Damaged composite laminates: assessment of residual Young's modulus through the impulse excitation technique. Compos B 128:76–82

Pintelon R, Schoukens J (2012) System identification: a frequency domain approach. Wiley, Hoboken, NJ

Pinter SZ, Fernando XN (2010) Estimation and equalization of fiber-wireless uplink for multiuser CDMA 4G networks. IEEE Trans Commun 58:1803–1813

Roinila T, Helin T, Vilkko M, Suntio T, Koivisto H (2009) Circular correlation based identification of switching power converter with uncertainty analysis using fuzzy density approach. Simul Model Pract Theor 17:1043–1058

Schoukens J, Pintelon R, Rolain Y (2012) Mastering system identification in 100 exercises. Wiley, Hoboken, NJ

Schoukens J, Godfrey K, Schoukens M (2018) Nonparametric data-driven modeling of linear systems. IEEE Control Syst Mag 38:49–88

Tan AH (2018) Multi-input identification using uncorrelated signals and its application to dual-stage hard disk drives. IEEE Trans Magn 54: Article 9300604

Tan AH, Godfrey KR (2004) Modeling of direction-dependent processes using Wiener models and neural networks with nonlinear output error structure. IEEE Trans Instrum Meas 53:744–753

Tan AH, Barker HA, Godfrey KR (2015) Identification of multi-input systems using simultaneous perturbation by pseudorandom input signals. IET Control Theory Appl 9:2283–2292

Visacro S, Guimarães MB, Araujo LS (2013) Experimental impulse response of grounding grids. Electr Power Syst Res 94:92–98

Xue S, Yang X, Li Z, Gao H (2018) An approach to fault detection for multirate sampled-data systems with frequency specifications. IEEE Trans Syst, Man, and Cybern: Syst 48:1155–1165

Yang C-C (2008) Optical CDMA fiber radio networks using cyclic ternary sequences. IEEE Commun Lett 12:41–43

Chapter 2
Design of Pseudorandom Signals for Linear System Identification

2.1 Maximum Length Binary Signals

The most common form of PRB signal is the MLB signal which is generated in GF(2). The MLB signals exist for lengths $N = 2^n - 1$, where n is an integer >1 so that $N = 3, 7, 15, 31, 63, 127, 255, 511, 1023, 2047, \ldots$. For each value of $N \geq 7$, there is more than one different MLB signal. MLB signals can be generated in hardware using shift registers consisting of n stages, which is an important advantage (Golomb 2017); this partly leads to their popularity in the industry. The feedback to the first stage is the modulo-2 sum of the logic value of the last stage and one or more of the other stages. For a particular feedback connection to result in an MLB signal, the corresponding characteristic equation in the delays D in the shift register must be irreducible and primitive. A polynomial of degree n is irreducible if it is not divisible by any polynomial of degree less than n but greater than zero. This polynomial is primitive if and only if it does not divide $x^r \oplus_2 1$ for any $r < 2^n - 1$, where $\oplus_2$ represents modulo-2 addition (Godfrey 1993; Tan and Godfrey 2002).

If the polynomial $c_n x^n \oplus_2 c_{n-1} x^{n-1} \oplus_2 \ldots \oplus_2 c_1 x \oplus_2 c_0 = 0$ is primitive, the characteristic equation in the delays introduced by the shift register is

$$c_n D^n \oplus_2 c_{n-1} D^{n-1} \oplus_2 \ldots \oplus_2 c_1 D \oplus_2 c_0 = 0. \tag{2.1}$$

Using the fact that modulo-2 addition is equivalent to modulo-2 subtraction, the feedback configuration is given by

$$c_0 X = c_1 DX \oplus_2 \ldots \oplus_2 c_{n-1} D^{n-1} X \oplus_2 c_n D^n X, \tag{2.2}$$

where X is the input to the shift register, DX is the sequence at the output of the first stage of the register and so on, so that $D^n X$ is the sequence at the output of the last stage of the n-stage register. The initial values stored in the register may be any

A. H. Tan and K. R. Godfrey, *Industrial Process Identification*, Advances in Industrial Control, https://doi.org/10.1007/978-3-030-03661-4_2

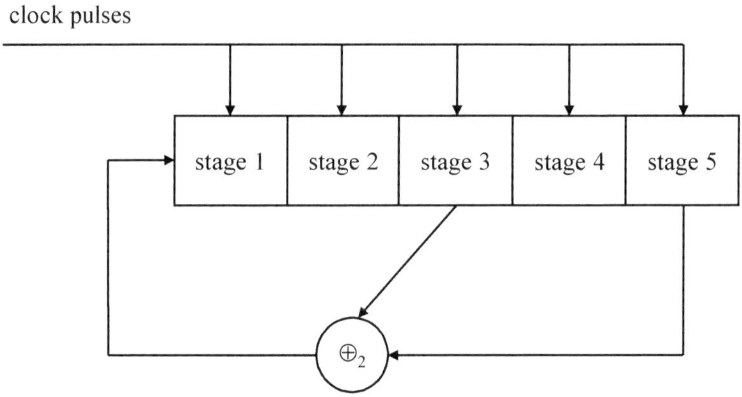

Fig. 2.1 Shift register for generating an MLB signal with $N = 31$

Table 2.1 Computation of the first 16 bits of the MLB signal with characteristic equation $D^5 \oplus_2 D^3 \oplus_2 1 = 0$

Clock pulse	Value in stage 1	Value in stage 2	Value in stage 3	Value in stage 4	Value in stage 5	Output (signal level)
1 (initialisation)	1	0	0	0	0	−1
2	0	1	0	0	0	−1
3	0	0	1	0	0	−1
4	1	0	0	1	0	−1
5	0	1	0	0	1	1
6	1	0	1	0	0	−1
7	1	1	0	1	0	−1
8	0	1	1	0	1	1
9	0	0	1	1	0	−1
10	1	0	0	1	1	1
11	1	1	0	0	1	1
12	1	1	1	0	0	−1
13	1	1	1	1	0	−1
14	1	1	1	1	1	1
15	0	1	1	1	1	1
16	0	0	1	1	1	1

sequence consisting of bits 0 and 1, except all 0's. An MLB sequence is obtained by reading the value at any of the stages of the shift register. This sequence is then converted into an MLB signal by converting the bits with value 1 to signal level 1 and the bits with value 0 to signal level −1, or vice versa. The signals may be scaled to any amplitude in a straightforward manner but here, it is assumed that the amplitudes are ±1 for the sake of simplicity.

As an example, the shift register corresponding to the characteristic equation $D^5 \oplus_2 D^3 \oplus_2 1 = 0$ is shown in Fig. 2.1. The signal has a period of $N = 31$. The computation of the first 16 bits of the MLB signal is illustrated in Table 2.1. The shift register is initialised with starting bits of [1 0 0 0 0]. The input to stage 1 is calculated by taking (value in stage 3) $\oplus_2$ (value in stage 5). The output is mapped from the value in stage 5 through a sequence-to-signal conversion.

The MLB signal has the following properties:

- The signal is binary, with levels +1 and −1.
- The RMS value is 1.
- The jth moment, $M_j \equiv E[u^j(i)]$, $j \in Z^+$, is $1/N$ for j odd and 1 for j even. See Eq. 5.4 for the definition of moment.
- The DFT magnitude is given by

$$|U(k)| = \begin{cases} 1 & k = 0 \\ \sqrt{N+1} & k \neq 0 \end{cases}, \tag{2.3}$$

  where all the harmonics have equal DFT magnitude except harmonic 0.
- The normalised autocorrelation function, $R_{uu}(n) = \frac{1}{N}\sum_{i=1}^{N} u(i)u(i+n)$, is given by

$$R_{uu}(n) = \begin{cases} 1 & n = 0 \\ -1/N & n \neq 0 \end{cases}. \tag{2.4}$$

  The autocorrelation function resembles an impulse function which gives the MLB signal pseudo-noise properties. This results in the MLB signal being utilised in many applications which require a periodic signal with noise-like properties. Some recent works are reported by Debenjak et al. (2014), Neshvad et al. (2015) and Davidson et al. (2015).
- The signal has a PIPS value of

$$\text{PIPS} = 100\sqrt{(N^2 - 1)/N^2}\%, \tag{2.5}$$

  which tends to 100% as N increases.
- Due to the effect of ZOH, the highest harmonic R to be used in the experiment has the least power. If $R = (N-1)/2$, PIPSE is maximised and for a uniform harmonic specification, PIPSE approaches 88.0%, while EMINE approaches 72.4% as N increases (Godfrey et al. 2005).
- The signal possesses the shift-and-multiply property such that

$$u(i-\alpha)u(i-\beta) = -u(i-\gamma), \tag{2.6}$$

  where α, β and γ are integers. (For the MLB sequence with values 0 and 1, the shift-and-multiply property becomes the shift-and-add property.) Thus, terms with multiple products of inputs u can be replaced by terms with single lag in u. These lags are dependent on the particular MLB signal used and are different for different MLB signals of the same period. They can be applied for the detection of the presence of certain types of nonlinear distortion, as will be illustrated in Chap. 6.

An example of the MLB signal with characteristic equation $D^5 \oplus_2 D^3 \oplus_2 1 = 0$ is shown in Fig. 2.2.

An inverse-repeat MLB signal can be generated by concatenating two periods of an MLB signal and inverting every other bit, giving a new signal with period twice that of the original MLB signal. For example, the characteristic equation $D^3 \oplus_2 D^2 \oplus_2 1 = 0$ generates an MLB signal $[-1\ -1\ 1\ -1\ 1\ 1\ 1]$ with period $N = 7$. The original MLB signal is concatenated with itself to give $[-1\ -1\ 1\ -1\ 1\ 1\ 1\ -1\ -1\ 1\ -1\ 1\ 1\ 1]$. Every other bit of it is then inverted, resulting in an inverse-repeat MLB signal of period $N = 14$ given by $[-1\ 1\ 1\ 1\ 1\ -1\ 1\ 1\ -1 -1\ -1\ -1\ 1\ -1]$. The second half of the inverse-repeat signal is the negative of the first half. This property ensures that the signal only contains power at the odd harmonics.

The inverse-repeat MLB signal has the following properties:

- The signal is binary, with levels +1 and -1.
- The RMS value is 1.
- The jth moment, $\mathrm{M}_j \equiv \mathrm{E}[u^j(i)]$, $j \in \mathrm{Z}^+$, is 0 for j odd and 1 for j even.
- The DFT magnitude is given by

$$|U(k)| = \begin{cases} 0 & k \in \{2p | p \in \mathrm{Z}\} \\ \sqrt{2N+4} & k \in \{1+2p | p \in \mathrm{Z}\}, k \notin N/2 \\ 2 & k = N/2 \end{cases}. \quad (2.7)$$

  The fact that even harmonics have zero DFT magnitude enables the signal to be applied in situations where the effects of even-order nonlinearities are to be eliminated. Examples of recent work using the inverse-repeat MLB signal are given by Roinila et al. (2014) and Nguyen et al. (2018). The signal has equal DFT magnitude at all the odd harmonics except the Nyquist frequency.
- The normalised autocorrelation function, $R_{uu}(n) = \frac{1}{N}\sum_{i=1}^{N} u(i)u(i+n)$, is given by

$$R_{uu}(n) = \begin{cases} 1 & n = 0 \\ -1 & n = N/2 \\ 2/N & n \in \{1+2p | p \in \mathrm{Z}\}, n \notin N/2 \\ -2/N & n \in \{2p | p \in \mathrm{Z}\}, n \notin 0 \end{cases}. \quad (2.8)$$

- The signal has a PIPS value of 100%.
- The highest harmonic R to be used in the experiment has the least power due to the effect of the ZOH. If $R = (N-2)/2$, PIPSE is maximised. (This assumes that the

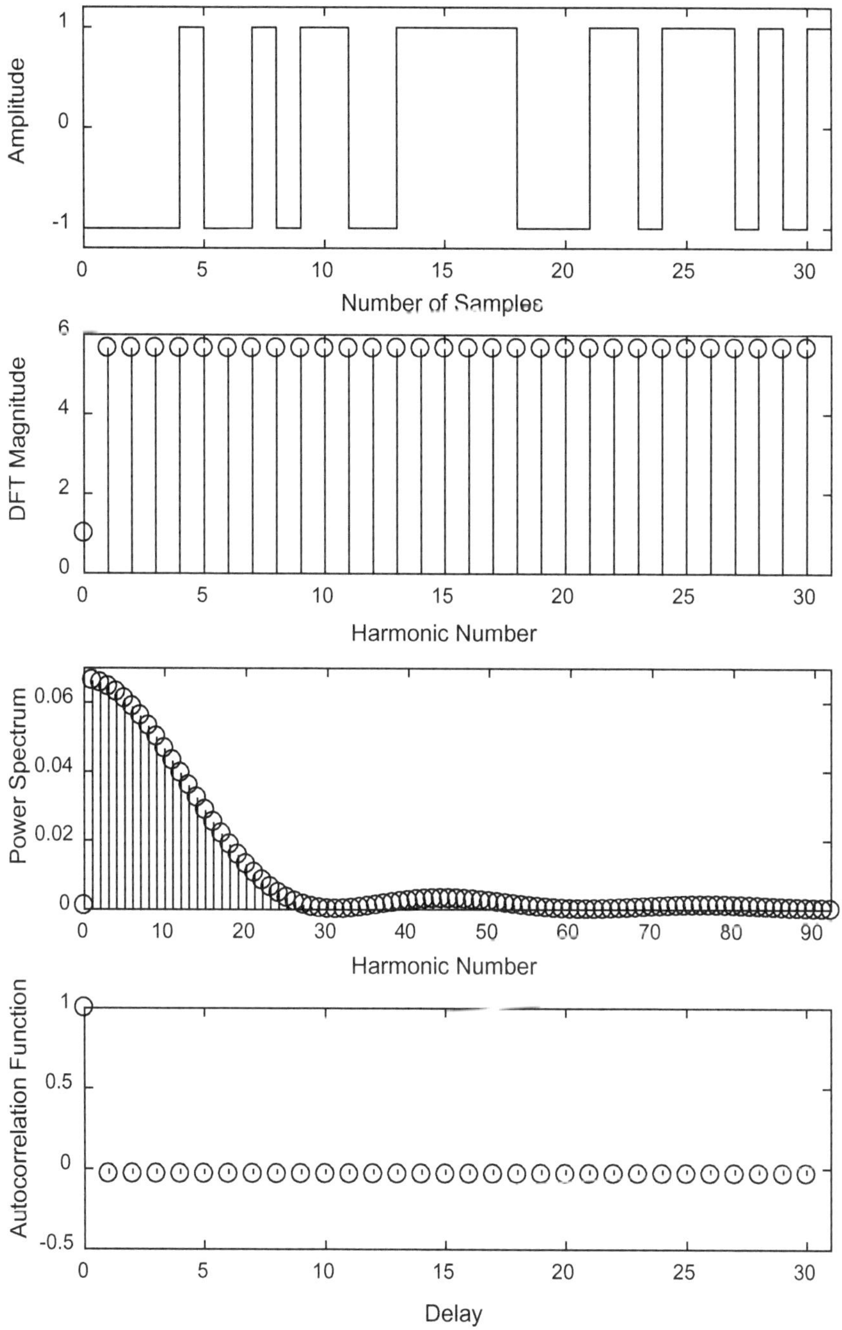

Fig. 2.2 An MLB signal of period $N = 31$

Nyquist frequency is not used, as is the normal practice in identification.) With R set at this value, then for a uniform harmonic specification for the odd harmonics, as N increases, PIPSE and EMINE tend to 88.0 and 72.4%, respectively.

- The signal possesses the shift-and-multiply property only for odd number of terms. For example,

$$u(i-\alpha)u(i-\beta)u(i-\chi)=u(i-\gamma), \tag{2.9}$$

where α, β, χ and γ are integers. However, the multiplication of an even number of terms such as $u(i-\alpha)u(i-\beta)$ will not result in a shifted version of the signal.

An example of the inverse-repeat MLB signal with characteristic equation $D^5 \oplus_2 D^3 \oplus_2 1 = 0$ is shown in Fig. 2.3.

2.2 Other Binary and Near-Binary Signals

Although the MLB signal is the most well-known type of PRB signal, several other types of PRB signal exist. These are

1. QRB signals (Charters 2009; Egidi and Manzini 2013), for which $N = 4k-1$ and prime ($N = 3, 7, 11, 19, 23, 31, 43, 47, 59, 67, 71, 79, 83, 103, 107, 127,\ldots$),
2. HAB (also known as Hall's sextic residue) signals (Lee et al. 2015), for which $N = 4k^2 + 27$ and prime ($N = 31, 43, 127, 223, 283, 811, 1051, 1471, 1627,\ldots$), and
3. TPB signals (Dai et al. 2009), for which $N = k(k + 2)$, with both k and $(k + 2)$ prime ($N = 15, 35, 143, 323, 899, 1763, 3599, 5183,\ldots$).

These signals are useful for filling the gaps between the available periods of the MLB signal. The formulae for generating these signals can be found in Everett (1966) and Tan and Godfrey (2002).

The QRB, HAB and TPB signals share the same properties as MLB signals except that they generally do not possess the shift-and-multiply property, unless the signals generated coincide with the MLB signals. They can also be made inverse-repeat, by concatenating two periods of the original signal and inverting every other bit of the new signal with period twice that of the original signal. The inverse-repeat signal shares the same properties as the inverse-repeat MLB signals, except that the shift-and-multiply property does not generally apply.

2.2.1 *Quadratic Residue Binary Signals*

The QRB signal $u(i)$, $i = 1, 2,\ldots, N$ is formed from the rule

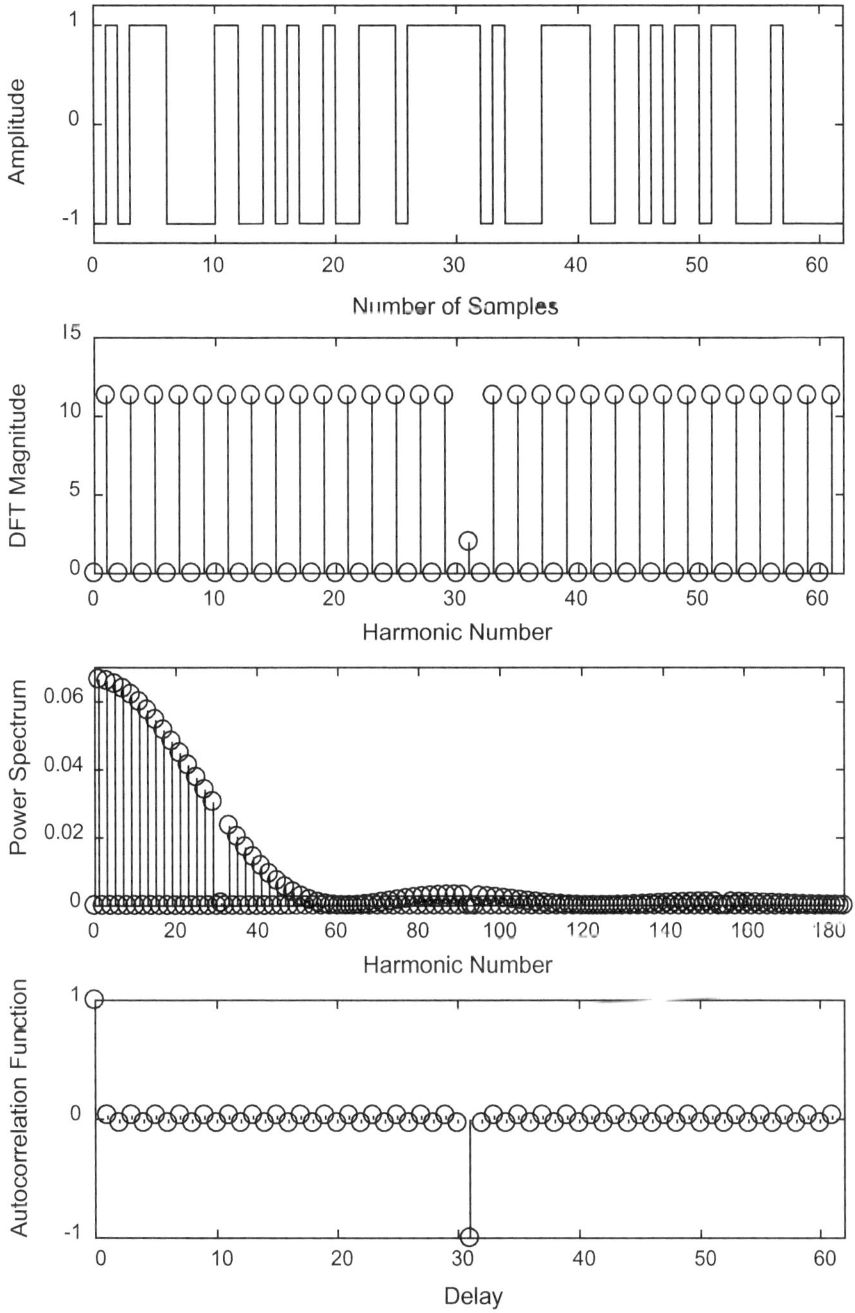

Fig. 2.3 An inverse-repeat MLB signal of period $N = 62$

$$\begin{aligned} u(i) &= 1 \text{ if } i \text{ is a square, modulo-}N \\ u(i) &= -1 \text{ otherwise} \\ u(N) &= 1 \text{ or } -1 \end{aligned} \quad . \tag{2.10}$$

For example, to generate a QRB signal with $N = 11$, first, form the sequence [1 2 3 … $(N-1)/2$] = [1 2 3 4 5]. Square this sequence to give [1 4 9 16 25]. Then take the modulo-11 of the squared sequence to give [1 4 9 5 3]. Now set $u(1) = u(3) = u(4) = u(5) = u(9) = 1$. Assume that $u(11)$ is taken as 1. (It can be equally well taken as -1.) The rest of the bits are set to -1. The QRB signal is [1 -1 1 1 1 -1 -1 -1 1 -1 1].

2.2.2 Hall Binary Signals

To generate a HAB signal, a primitive root r of N is first chosen such that if $3 \equiv r^t$ (modulo-N), then $t \equiv 1$ (modulo-6). The signal is formed from the rule that

$$\begin{aligned} u(i) &= 1 \text{ if } i \equiv r^s, \text{ modulo-}N, \text{ where } s \equiv 0,\ 1 \text{ or } 3 \text{ (modulo-6)} \\ u(i) &= -1 \text{ otherwise} \end{aligned} \quad . \tag{2.11}$$

For example, to generate a HAB signal with $N = 31$, a primitive root is selected as 3. Form the sequence $r^s = [\,3^0\ 3^1\ 3^3\ 3^6\ 3^7\ 3^9\ 3^{12}\ 3^{13}\ 3^{15}\ 3^{18}\ 3^{19}\ 3^{21}\ 3^{24}\ 3^{25}\ 3^{27}\ 3^{30}\,]$ where the powers are the values of s that satisfy $s \equiv 0,\ 1$ or 3 (modulo-6). Obtain its remainder modulo-31. This results in [1 3 27 16 17 29 8 24 30 4 12 15 2 6 23 1]. Now set $u(1) = u(3) = u(27) = u(16) = u(17) = u(29) = u(8) = u(24) = u(30) = u(4) = u(12) = u(15) = u(2) = u(6) = u(23) = 1$. The rest of the bits are set to -1.

2.2.3 Twin Prime Binary Signals

TPB signals can be formed from QRB signals. First QRB signals are generated for both periods p and $p + 2$; these signals are denoted by $a(i)$ and $b(i)$, respectively. The TPB signal $u(i)$ is then defined by

$$\begin{aligned} u(i) &= a(i)b(i) \text{ if } i \neq 0 \text{ modulo-}p \text{ or modulo-}(p+2) \\ u(i) &= 1 \text{ if } i = 0 \text{ modulo-}(p+2) \\ u(i) &= -1 \text{ if } i = 0 \text{ modulo-}p, \text{ but } i \neq 0 \text{ modulo-}(p+2) \end{aligned} \quad . \tag{2.12}$$

For example, to generate a TPB signal with $N = 15$, QRB signals of periods 3 and 5 are first generated giving the signals [1 -1 1] and [1 -1 -1 1 1], respectively.

These are concatenated 5 times and 3 times, respectively, to result in signals $a(i)=$ [1 −1 1 1 −1 1 1 −1 1 1 −1 1 1 −1 1] and $b(i)=$[1 −1 −1 1 1 1 −1 −1 1 1 1 −1 −1 1 1]. From Eq. 2.12, $u(i)=1$ for $i=5$, 10 and 15, and $u(i)=-1$ for $i=3$, 6, 9 and 12. Finally, $u(i)=a(i)b(i)$ for $i=1$, 2, 4, 7, 8, 11, 13 and 14. Hence, $u(1)=1$, $u(2)=1$, $u(4)=1$, $u(7)=-1$, $u(8)=1$, $u(11)=-1$, $u(13)=-1$ and $u(14)=-1$.

2.2.4 Quadratic Residue Ternary Signals

The QRT signal is a near-binary signal which is closely related to the QRB signal but has $u(N)=0$ instead of +1 or −1. QRT signals exist for $N=4k\pm 1$, where k is an integer and N is prime, that is, $N=3$, 5, 7, 11, 13, 17, 19, 23, 29, 31, 37, 41, 43,…. Note that more periods are available than for other types of pseudorandom signal, and hence it is a useful addition to the PRB signals mentioned above. This is particularly so as the QRT signal possesses properties which are very close to those of the PRB signals. The power in the signal is slightly lower for the QRT signal compared with the PRB signals due to 1 bit being at 0 in a period N.

The QRT signal $u(i)$, $i=1, 2,\ldots, N$ is formed from the rule

$$\begin{aligned} u(i) &= 1 \text{ if } i \text{ is a square, modulo-}N \\ u(i) &= -1 \text{ otherwise} \\ u(N) &= 0 \end{aligned} \quad . \tag{2.13}$$

The QRT signal has the following properties:

- The signal is ternary, with levels +1, −1 and 0. In a period N, there are $(N-1)/2$ bits at 1, $(N-1)/2$ bits at −1 and 1 bit at 0.
- The RMS value is $\sqrt{\frac{N-1}{N}}$.
- The jth moment, $M_j \equiv E[u^j(i)]$, $j \in Z^+$, is 0 for j odd and $\frac{N-1}{N}$ for j even.
- The DFT magnitude is given by

$$|U(k)| = \begin{cases} 0 & k=0 \\ \sqrt{N} & k \neq 0 \end{cases} \tag{2.14}$$

  where all the harmonics have equal DFT magnitude except harmonic 0.
- The normalised autocorrelation function, $R_{uu}(n)=\frac{1}{N}\sum_{i=1}^{N}u(i)u(i+n)$, is given by

$$R_{uu}(n) = \begin{cases} (N-1)/N & n=0 \\ -1/N & n\neq 0 \end{cases} . \tag{2.15}$$

- The signal has a PIPS value of

$$\text{PIPS} = 100\sqrt{(N-1)/N}\% \tag{2.16}$$

which tends to 100% as N increases.

An example of a QRT signal with $N = 11$ is shown in Fig. 2.4, where it is can be seen that the signal is near-binary with a single bit at signal level 0. This becomes more apparent as N increases.

The QRT signal can also be made inverse-repeat, by concatenating two periods of the original signal and inverting every other bit of the new signal with period twice that of the original signal. The inverse-repeat QRT signal shares properties which are very close to those of the inverse-repeat PRB signals.

The inverse-repeat QRT signal has the following properties:

- The signal is ternary, with levels +1, −1 and 0. In a period N, there are $(N-2)/2$ bits at 1, $(N-2)/2$ bits at −1 and 2 bits at 0.
- The RMS value is $\sqrt{\frac{N-2}{N}}$.
- The jth moment, $M_j \equiv E[u^j(i)]$, $j \in Z^+$, is 0 for j odd and $\frac{N-2}{N}$ for j even.
- The DFT magnitude is given by

$$|U(k)| = \begin{cases} 0 & k \in \{2p|p \in Z\} \\ \sqrt{2N} & k \in \{1+2p|p \in Z\}, k \notin N/2 \\ 0 & k = N/2 \end{cases} \tag{2.17}$$

  where all the odd harmonics have equal DFT magnitude except the Nyquist frequency.
- The normalised autocorrelation function, $R_{uu}(n) = \frac{1}{N}\sum_{i=1}^{N} u(i)u(i+n)$, is given by

$$R_{uu}(n) = \begin{cases} (N-2)/N & n = 0 \\ -(N-2)/N & n = N/2 \\ 2/N & n \in \{1+2p|p \in Z\}, n \notin N/2 \\ -2/N & n \in \{2p|p \in Z\}, n \notin 0 \end{cases}. \tag{2.18}$$

- The signal has a PIPS value of

$$\text{PIPS} = 100\sqrt{(N-2)/N}\% \tag{2.19}$$

  which tends to 100% as N increases.

An example of an inverse-repeat QRT signal with $N = 22$ is shown in Fig. 2.5. The signal is generated from a QRT signal with an original period of 11 which is shown in Fig. 2.4.

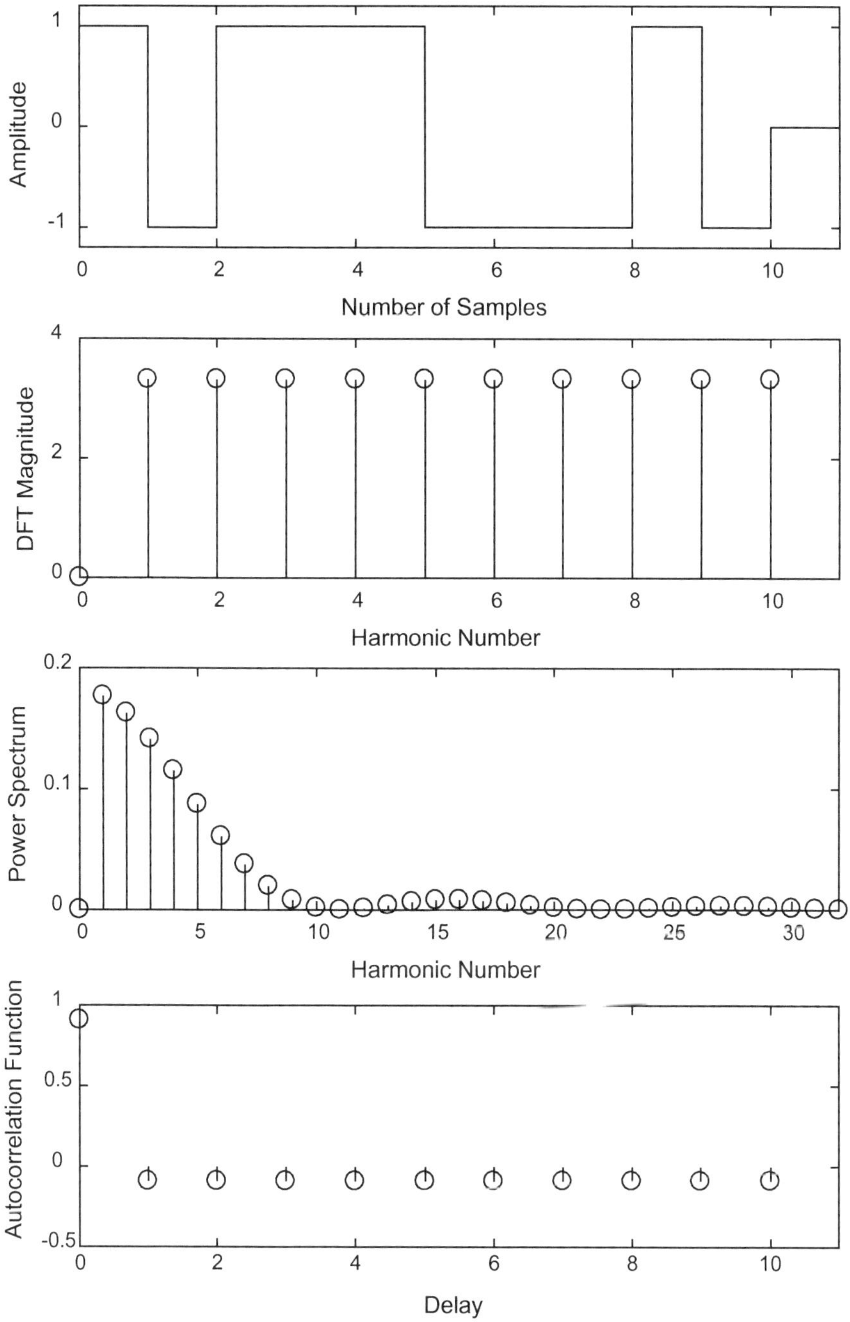

Fig. 2.4 A QRT signal of period $N = 11$

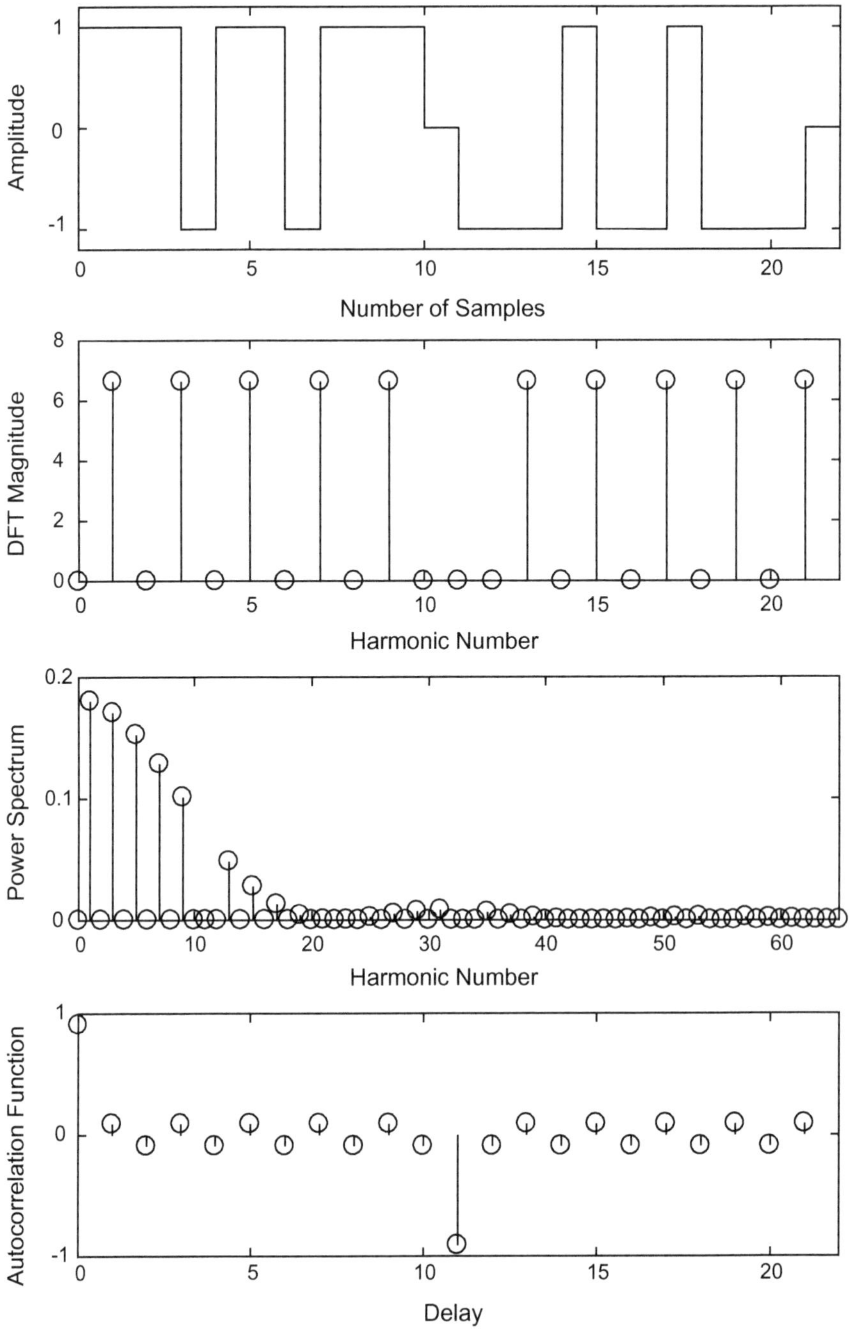

Fig. 2.5 An inverse-repeat QRT signal of period $N = 22$

2.3 Multilevel Maximum Length Signals

2.3.1 Pseudorandom Signals with Maximum Length

Multilevel maximum length signals, or generally known as PRML signals, are based on maximum length sequences in GF(q), where q is a prime or a power of a prime. A PRML sequence $s_{q,n}$ (i) can be generated using a shift register consisting of n stages with modulo-q operations. The recurrence relation (Barker 1993) is given by

$$s_{q,n}(i) = -\sum_{r=1}^{n} c_r s_{q,n}(i - r), \text{ all } i \tag{2.20}$$

where c_r are the coefficients of a primitive polynomial of degree n in GF(q). Note that

$$(-1)^n c_n = g \tag{2.21}$$

is a primitive element of the field. A primitive element g has the important property that its powers g^r, for $r = 0, 1, \ldots, q - 2$, generate all the nonzero elements of the field.

The sequence $s_{q,n}(i)$ has period $N = q^n - 1$, where n is a positive integer. In a period, the field element 0 occurs $q^{n-1} - 1$ times while each of the elements 1, 2, …, $q - 1$ occurs q^{n-1} times. A PRML signal $u_{q,n}(i)$ can be obtained from $s_{q,n}(i)$ by converting each element 0, 1, 2, …, $q - 1$ of $s_{q,n}(i)$ into a signal level via a sequence-to-signal conversion. The resulting signal has the same period $N = q^n - 1$. When $q = 2$, the signals are simply MLB signals. When $q > 2$ and odd, the field elements can be converted into any odd number of signal levels from 3 to q. The harmonic properties are determined by the sequence-to-signal conversion. When $q > 2$ and odd, it is possible to generate PRML signals in which even harmonics are suppressed and odd harmonics are uniform (Barker and Zhuang 1997).

Harmonic multiples of both two and three can be suppressed provided the number of signal levels is greater than two and $q - 1$ is an integer multiple of six, for example, when $q = 7$, 13, 19, 25 and 31. This is because $q - 1$ must be an integer multiple of both two and three in this case. For $q = 31$, it is also possible to suppress harmonic multiples of five, since $q - 1$ is also an integer multiple of five. The available periods N for $q = 3$, 5, 7, 9, 11 and 13 are shown in Table 2.2 for n from 1 to 5. The gaps between these periods become increasingly large as n increases.

An example of a 7-level PRML signal from GF(7) with characteristic equation $3D^2 \oplus_7 D \oplus_7 1 = 0$ is shown in Fig. 2.6. The signal period is $N = 48$. The signal has harmonic multiples of two and three suppressed. The sequence-to-signal conversion is shown in Table 2.3. The signal has a PIPS value of 67.36%. A higher PIPS value may be obtained if a smaller number of signal levels are used. For a 3-level signal from GF(7) with harmonic multiples of two and three suppressed, a PIPS value of

Table 2.2 The periods N for $q = 3, 5, 7, 9, 11$ and 13 for n from 1 to 5

q	$n = 1$	$n = 2$	$n = 3$	$n = 4$	$n = 5$
3	2	8	26	80	242
5	4	24	124	624	3124
7	6	48	342	2400	16,806
9	8	80	728	6560	59,048
11	10	120	1330	14,640	161,050
13	12	168	2196	28,560	371,292

Table 2.3 Sequence-to-signal conversion for a 7-level PRML signal from GF(7)

Sequence value (field element)	0	1	2	3	4	5	6
Signal level	0	1	2	3	−3	−2	−1

Table 2.4 Sequence-to-signal conversion for a 3-level PRML signal from GF(7)

Sequence value (field element)	0	1	2	3	4	5	6
Signal level	0	1	0	1	−1	0	−1

76.38% can be achieved. The sequence-to-signal conversion is shown in Table 2.4, and the signal is plotted in Fig. 2.7. The power in the 3-level signal is lower than that in the 7-level signal as can be seen from the DFT magnitude and power spectrum plots. This is simply due to the fact that the maximum amplitude is larger for the 7-level signal.

2.3.2 *Truncated Pseudorandom Signals*

Truncated PRML signals have periods which are shorter than $N = q^n - 1$. These signals are very useful for filling up the gap between the available periods of the PRML signals which become increasingly large for large q and n, as can be seen from Table 2.2.

The design of truncated PRML signals relies on the design of the primitive signals $u_{q,1}(i)$, when $n = 1$. The primitive signal has DFT magnitude which is closely related to those of PRML signals with $n \neq 1$ (Tan et al. 2005). In particular,

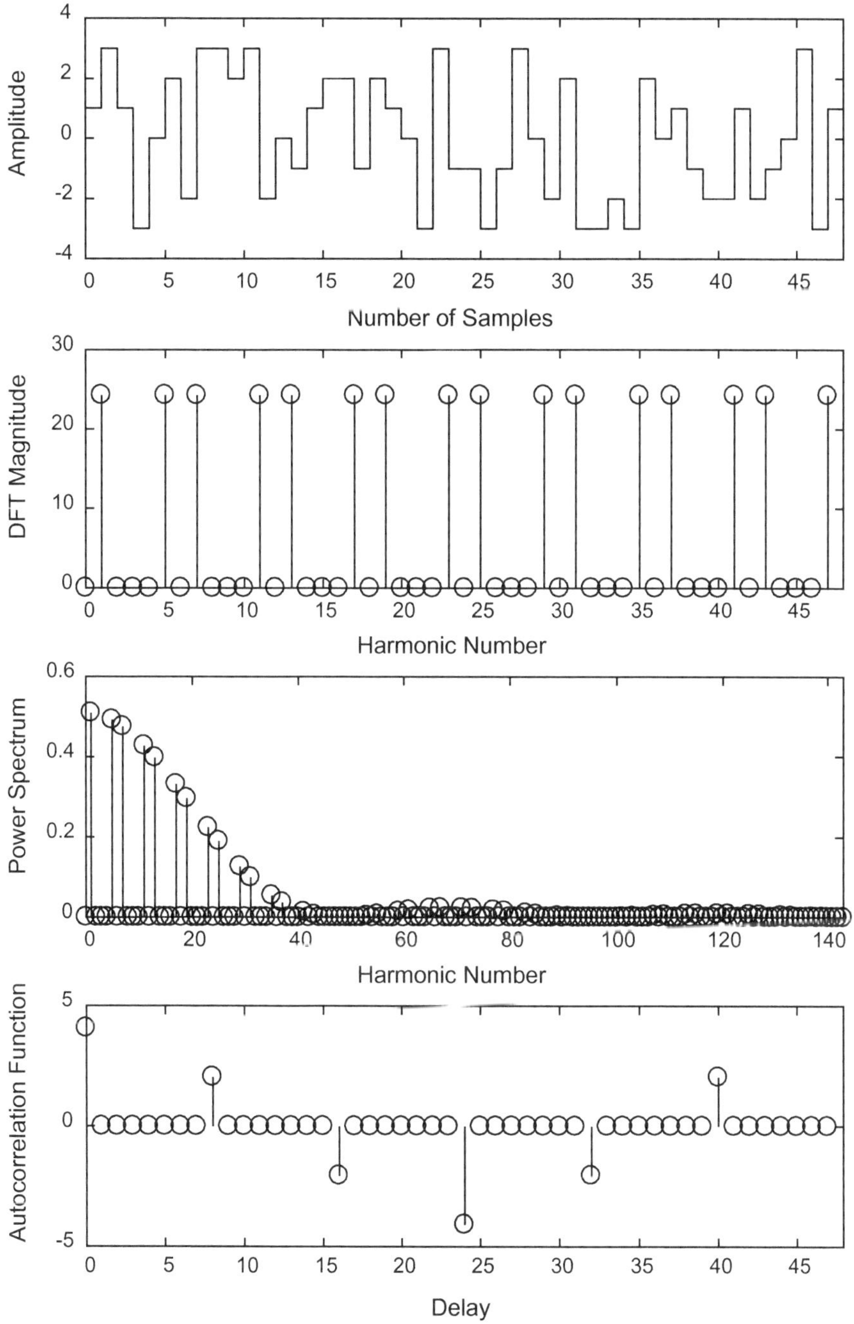

Fig. 2.6 A 7-level PRML signal from GF(7) with characteristic equation $3D^2 \oplus_7 D \oplus_7 1 = 0$ and $N = 48$

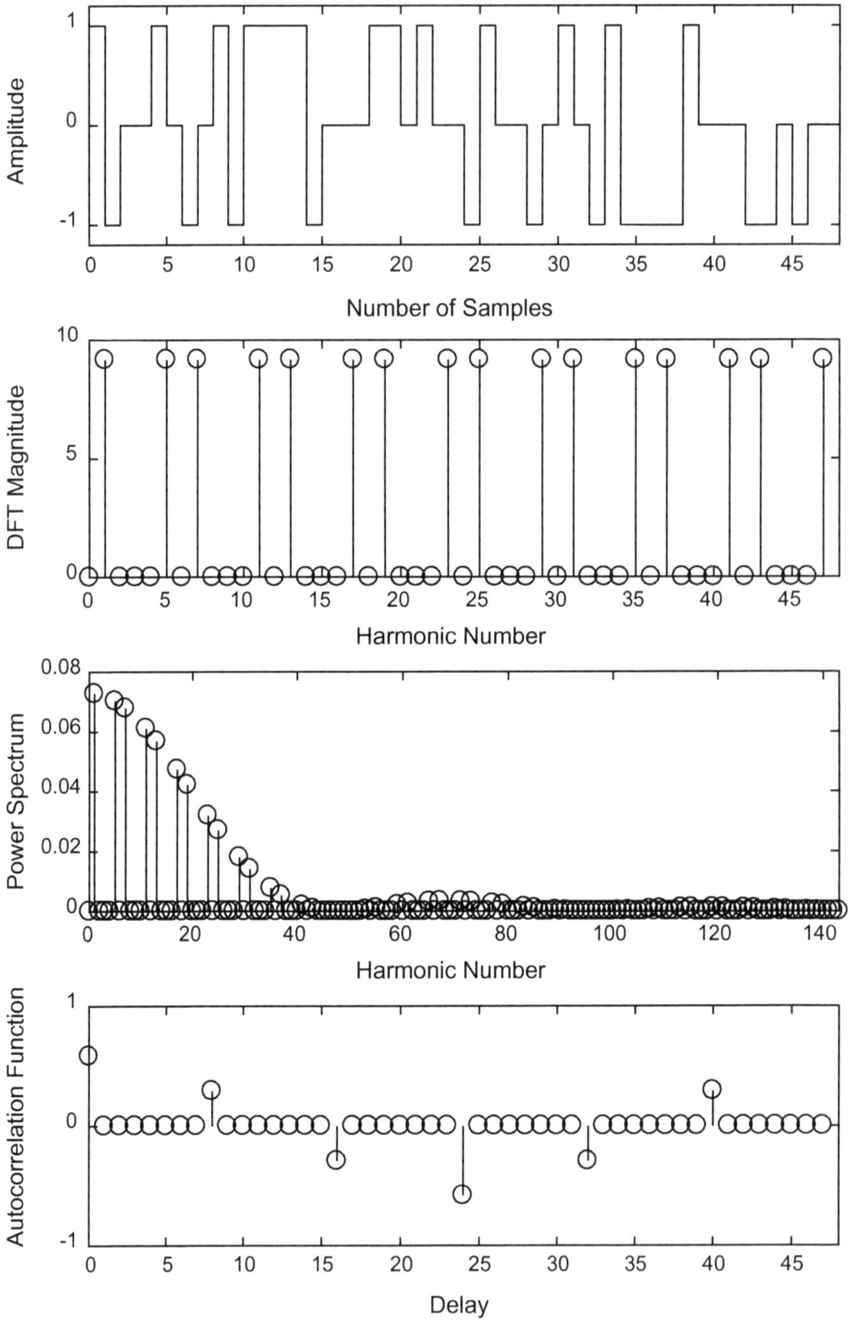

Fig. 2.7 A 3-level PRML signal from GF(7) with characteristic equation $3D^2 \oplus_7 D \oplus_7 1 = 0$ and $N = 48$

$$|U_{q,n}(k)| = \begin{cases} |U_{q,1}(0)|q^{n-1} & k \text{ a multiple of } q^n - 1 \\ |U_{q,1}(0)|q^{(n-2)/2} & k \text{ a multiple of } q - 1 \text{ but not of } q^n - 1 \\ |U_{q,1}(k)|q^{(n-1)/2} & k \text{ not a multiple of } q - 1 \end{cases}. \quad (2.22)$$

This means that the design of PRML signals can be done through the design of the corresponding primitive signal. If the primitive signal is designed to have subperiods within a period of $q - 1$, the PRML signal with $n \neq 1$ will also have subperiods within $q^n - 1$. Thus, the resulting signal has a period shorter than $q^n - 1$ and the signal is known as a truncated PRML signal.

While many possibilities exist in terms of the design of the primitive signal, two special choices lead to truncated PRML signals with particularly useful properties:

- Design with even harmonics suppressed and odd harmonics uniform, with

$$u_{q,1}(i) = [\,1\ -1\ 1\ -1\ \ldots\ 1\ -1\,]. \quad (2.23)$$

  The signal $u_{q,1}(i)$ has $(q - 1)/2$ subperiods within $(q - 1)$. The resulting $u_{q,n}(i)$ with $n \neq 1$ has $(q - 1)/2$ subperiods within $(q^n - 1)$. Each subperiod of length $\frac{2(q^n-1)}{q-1}$ defines a single period of the truncated PRML signal.
- Design with harmonic multiples of two and three suppressed and the remaining harmonics uniform, with

$$u_{q,1}(i) = [\,1\ 1\ 0\ -1\ -1\ 0\ 1\ 1\ 0\ -1\ -1\ 0\ \ldots\ 1\ 1\ 0\ -1\ -1\ 0\,]. \quad (2.24)$$

  The signal $u_{q,1}(i)$ has $(q - 1)/6$ subperiods within $(q - 1)$. The resulting $u_{q,n}(i)$ with $n \neq 1$ has $(q - 1)/6$ subperiods within $(q^n - 1)$. Each subperiod of length $\frac{6(q^n-1)}{q-1}$ defines a single period of the truncated PRML signal. For this choice, $(q - 1)$ must be an integer multiple of six.

In generating the truncated PRML signals, $u_{q,1}(i)$ is used to define the sequence-to-signal conversion. An example is shown here for the design using Eq. 2.23 for $q = 7$. A primitive element of GF(7) is $g = 3$. From Eq. 2.21, $c_1 = -3$. The characteristic equation for $n = 1$ is $c_1 D \oplus_7 1 = 0$. The recurrence equation is $s_{7,1}(i) = -c_1 s_{7,1}(i - 1) = 3s_{7,1}(i - 1)$. Starting with the field element 1, the sequence $s_{7,1}(i) = [\,3^0\ 3^1\ 3^2\ 3^3\ 3^4\ 3^5\,]$ modulo-7 $= [\,1\ 3\ 2\ 6\ 4\ 5\,]$. Comparing with $u_{7,1}(i) = [\,1\ -1\ 1\ -1\ 1\ -1\,]$, the sequence-to-signal conversion is given in Table 2.5. An example of a truncated PRML signal designed using Eq. 2.23 is shown in Fig. 2.8. The signal is generated from GF(7) with characteristic equation $3D^2 \oplus_7 D \oplus_7 1 = 0$ and sequence-to-signal conversion as in Table 2.5. The signal has a PIPS value of 93.54%.

The truncated PRML signal designed using Eq. 2.23 has the following properties:

- The signal is ternary, with levels +1, −1 and 0.

Table 2.5 Sequence-to-signal conversion for a truncated PRML signal from GF(7) using Eq. 2.23

Sequence value (field element)	0	1	2	3	4	5	6
Signal level	0	1	1	−1	1	−1	−1

- The signal period is $N = \frac{2(q^n-1)}{q-1}$ (Tan 2007). In each period, each of the signal levels 1 and −1 occurs q^{n-1} times. The signal level zero occurs $\frac{2(q^{n-1}-1)}{q-1}$ times.
- The RMS value is $\sqrt{\frac{q^{n-1}(q-1)}{q^n-1}}$.
- The jth moment, $M_j \equiv E[u^j(i)]$, $j \in Z^+$, is 0 for j odd and $\left(\frac{q^{n-1}(q-1)}{q^n-1}\right)$ for j even.
- The DFT magnitude is given by (Tan 2007)

$$|U(k)| = \begin{cases} 2q^{(n-1)/2} & k \in \{1+2p|p \in Z\} \\ 0 & k \in \{2p|p \in Z\} \end{cases}. \tag{2.25}$$

  The squared version of the signal has uniform even harmonics which makes the signal very suitable for the identification of Hammerstein models with a linear pathway in parallel with a pathway consisting of a quadratic nonlinearity in series with a second linear block (Tan 2007).
- The normalised autocorrelation function, $R_{uu}(n) = \frac{1}{N}\sum_{i=1}^{N} u(i)u(i+n)$, is given by

$$R_{uu}(n) = \begin{cases} \frac{q^{n-1}(q-1)}{q^n-1} & n = 0 \\ -\frac{q^{n-1}(q-1)}{q^n-1} & n = N/2 \\ 0 & \text{otherwise} \end{cases}. \tag{2.26}$$

- The signal has a PIPS value of $100\sqrt{\frac{q^{n-1}(q-1)}{q^n-1}}\%$ (Tan 2007).

A list of some possible periods generated with $q \le 81$, as well as $\lim_{n\to\infty}$ PIPS, is shown in Table 2.6.

An example is shown here for the design using Eq. 2.24 for $q = 13$. A primitive element of GF(13) is $g = 2$. From Eq. 2.21, $c_1 = -2$. The characteristic equation for $n = 1$ is given by $c_1 D \oplus_{13} 1 = 0$. The recurrence equation is $s_{13,1}(i) = -c_1 s_{13,1}(i-1) = 2s_{13,1}(i-1)$. Starting with the field element 1, the sequence $s_{13,1}(i) = [\,2^0\ 2^1\ 2^2\ 2^3\ 2^4\ 2^5\ 2^6\ 2^7\ 2^8\ 2^9\ 2^{10}\ 2^{11}\,]$ modulo-13 $=$ [1 2 4 8 3 6 12 11 9 5 10 7]. Comparing with $u_{13,1}(i) =$ [1 1 0 −1 −1 0 1 1 0 −1 −1 0], the sequence-to-signal conversion is given Table 2.7. An example of a truncated PRML signal designed using Eq. 2.24 is

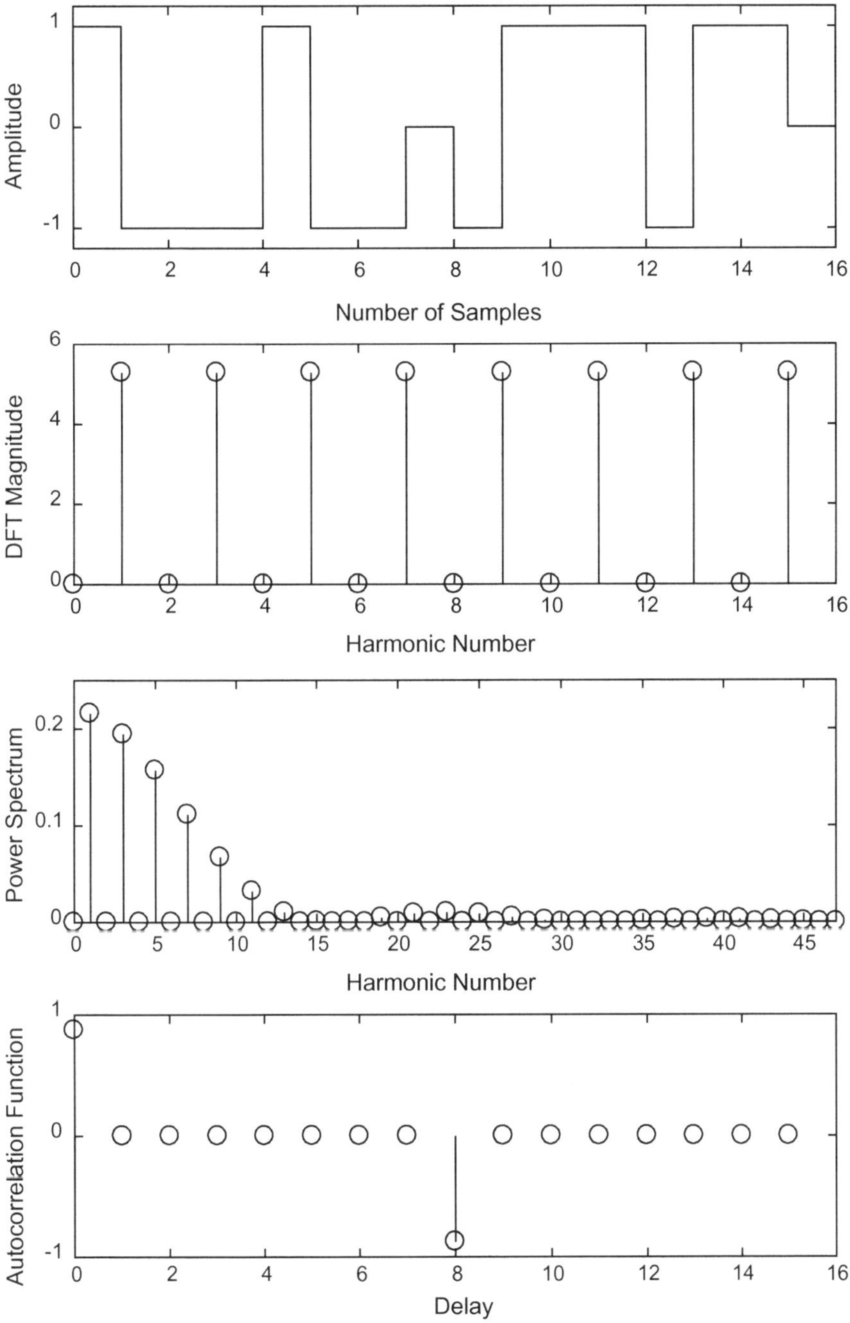

Fig. 2.8 A truncated PRML signal from GF(7) designed using Eq. 2.23, with characteristic equation $3D^2 \oplus_7 D \oplus_7 1 = 0$ and $N = 16$

Table 2.6 Possible periods of truncated PRML signals designed using Eq. 2.23 and generated with $q \leq 81$

q	Possible $N \leq 10{,}000$ where $n > 1$	$\lim_{n\to\infty}$ PIPS (%)
3	8, 26, 80, 242, 728, 2186, 6560	81.65
5	12, 62, 312, 1562, 7812	89.44
7	16, 114, 800, 5602	92.58
9	20, 182, 1640	94.28
11	24, 266, 2928	95.35
13	28, 366, 4760	96.08
17	36, 614	97.01
19	40, 762	97.33
23	48, 1106	97.80
25	52, 1302	97.98
27	56, 1514	98.13
29	60, 1742	98.26
31	64, 1986	98.37
37	76, 2814	98.64
41	84, 3446	98.77
43	88, 3786	98.83
47	96, 4514	98.93
49	100, 4902	98.97
53	108, 5726	99.05
59	120, 7082	99.15
61	124, 7566	99.18
67	136, 9114	99.25
71	144	99.29
73	148	99.31
79	160	99.37
81	164	99.38

shown in Fig. 2.9. The signal is generated from GF(13) with characteristic equation $2D^2 \oplus_{13} D \oplus_{13} 1 = 0$ and sequence-to-signal conversion as in Table 2.7. The signal has a PIPS value of 78.68%.

The truncated PRML signal designed using Eq. 2.24 has the following properties:

- The signal is ternary, with levels +1, −1 and 0.
- The signal period is $N = \frac{6(q^n - 1)}{q-1}$ (Tan and Foo 2006). In each period, each of the signal levels 1 and −1 occurs $2q^{n-1}$ times. The signal level zero occurs $2q^{n-1} + \frac{6(q^{n-1}-1)}{q-1}$ times.

Table 2.7 Sequence-to-signal conversion for a truncated PRML signal from GF(13) using Eq. 2.24

Sequence value (field element)	0	1	2	3	4	5	6	7	8	9	10	11	12
Signal level	0	1	1	−1	0	−1	0	0	−1	0	−1	1	1

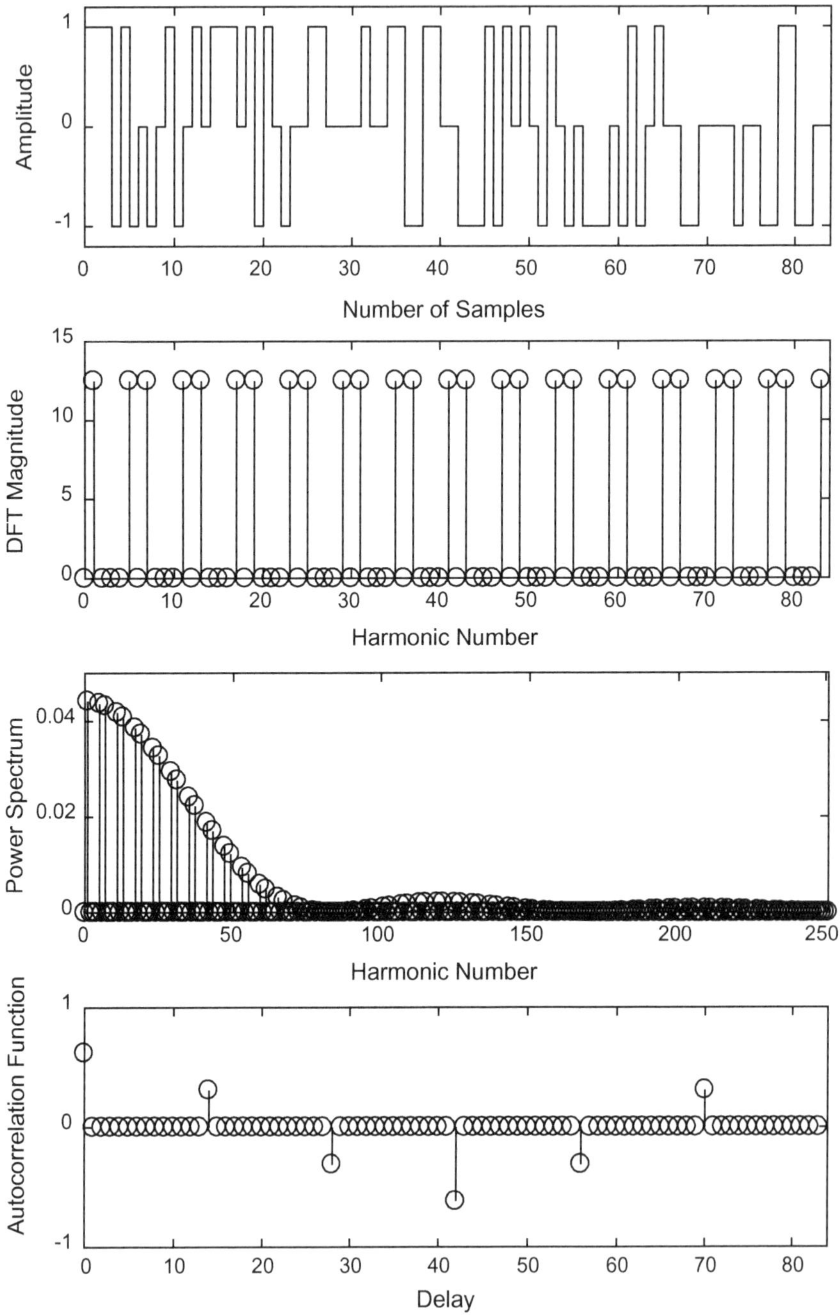

Fig. 2.9 A truncated PRML signal from GF(13) designed using Eq. 2.24, with characteristic equation $2D^2 \oplus_{13} D \oplus_{13} 1 = 0$ and $N = 84$

- The RMS value is $\sqrt{\frac{2q^{n-1}(q-1)}{3(q^n-1)}}$.
- The jth moment, $\mathrm{M}_j \equiv \mathrm{E}[u^j(i)]$, $j \in \mathrm{Z}^+$, is 0 for j odd and $\left(\frac{2q^{n-1}(q-1)}{3(q^n-1)}\right)$ for j even.
- The DFT magnitude is given by (Tan and Foo 2006)

$$|U(k)| = \begin{cases} 2\sqrt{3}q^{(n-1)/2} & k \in \{1+6p,\ 5+6p | p \in \mathrm{Z}\} \\ 0 & k \notin \{1+6p,\ 5+6p | p \in \mathrm{Z}\} \end{cases}. \tag{2.27}$$

- The normalised autocorrelation function, $R_{uu}(n) = \frac{1}{N}\sum_{i=1}^{N} u(i)u(i+n)$, is given by

$$R_{uu}(n) = \begin{cases} \frac{2q^{n-1}(q-1)}{3(q^n-1)} & n = 0 \\ -\frac{2q^{n-1}(q-1)}{3(q^n-1)} & n = N/2 \\ \frac{q^{n-1}(q-1)}{3(q^n-1)} & n = N/6, 5N/6 \\ -\frac{q^{n-1}(q-1)}{3(q^n-1)} & n = N/3, 2N/3 \\ 0 & n \notin Nm/6, m \in \mathrm{Z} \end{cases}. \tag{2.28}$$

- The signal has a PIPS value of $100\sqrt{\frac{2q^{n-1}(q-1)}{3(q^n-1)}}\%$ (Tan and Foo 2006).

A list of some possible periods generated with $q \leq 127$, as well as $\lim_{n\to\infty}$ PIPS, is given in Table 2.8. Some of these values can be found in Tan and Foo (2006).

2.4 Direct Synthesis Ternary Signals

Direct synthesis signals form a class of ternary signals with harmonic multiples of two and three suppressed. The motivation is first given for the design of such signals, followed by the theoretical derivation.

Consider a discrete ternary signal $u(i)$ having period N. In order to improve the SNR at the system output as well as the input magnitude uniformity at the excited frequencies, it is of interest to find $u(i)\forall i \in \{m\,|1 \leq m \leq N, m \in \mathrm{Z}^+\}$ in order to maximise the RMS value of the signal

$$\mathrm{rms}(u) = \sqrt{\frac{1}{N}\sum_{i=1}^{N} u^2(i)} \tag{2.29}$$

as well as the number of non-suppressed (excited) harmonics with uniform DFT magnitude represented by the cardinality of K, $\#K$, where $K = \{k|0 \leq k \leq N-1, k \notin P \cup Q, |U(k)| = \max(|U(k)|)\}$, subject to

Table 2.8 Possible periods of truncated PRML signals designed using Eq. 2.24 and generated with $q \leq 127$

q	Possible $N \leq 10{,}000$ where $n > 1$	$\lim_{n\to\infty}$ PIPS (%)
7	48, 342, 2400	75.59
13	84, 1098	78.45
19	120, 2286	79.47
25	156, 3906	80.00
31	192, 5958	80.32
37	228, 8442	80.54
43	264	80.69
49	300	80.81
61	372	80.98
67	408	81.04
73	444	81.09
79	480	81.13
97	588	81.23
103	624	81.25
109	660	81.27
121	732	81.31
127	768	81.33

$$U(k) = 0 \forall k \in P \cup Q, \tag{2.30}$$

and

$$u(i) \in \{1, 0, -1\}, \tag{2.31}$$

where $P = \{2m | 0 \leq m < N/2, m \in \mathrm{Z}\}$ and $Q = \{3m | 0 \leq m < N/3, m \in \mathrm{Z}\}$. This optimisation leads to the class of direct synthesis ternary signals, detailed in Tan (2013). Note that binary signals will not be able to satisfy the requirement of harmonic multiples of two and three suppressed; signals with at least three levels are necessary.

A signal $u(i)$ with period $N = 6N_{\text{basic}}$ satisfying Eq. 2.30 can be generated via direct synthesis using

$$u(i) = a(i) \times b(i), \tag{2.32}$$

provided

$$a(i) = [\, u_{\text{basic}\,(1)} \; u_{\text{basic}\,(2)} \; \cdots \; u_{\text{basic}\,(6)} \,], \tag{2.33}$$

$$b(i) = [\, u_{\text{special}\,(1)} \; u_{\text{special}\,(2)} \; \cdots \; u_{\text{special}(N_{\text{basic}})}], \tag{2.34}$$

where $u_{\mathrm{basic}(p)}(i)$ denotes the pth concatenation of $u_{\mathrm{basic}}(i)$. A similar notation is used for $u_{\mathrm{special}}(i)$. The signal $u_{\mathrm{basic}}(i)$ has period $N_{\mathrm{basic}} \in \{5+6p,\ 7+6p|p \geq 0,\ p \in \mathrm{Z}\} = \{5, 7, 11, 13, 17, 19, \ldots\}$. $N_{\mathrm{basic}} = 1$ is not considered here as it results in the trivial solution. In Eq. 2.34, the signal $u_{\mathrm{special}}(i)$ is designed to be the shortest possible ternary signal with harmonic multiples of two and three suppressed and with the excited harmonics having uniform DFT magnitude. This gives

$$u_{\mathrm{special}}(i) = \left[1\ 1\ 0\ -1\ -1\ 0\right]. \tag{2.35}$$

It is worth mentioning here that Eq. 2.32 can be applied to the design of signals with other harmonic properties which are not considered here, with some modifications to the signals $a(i)$ and $b(i)$.

If $u_{\mathrm{basic}}(i)$ has uniform DFT magnitude except at harmonic 0, such that

$$|U_{\mathrm{basic}}(k)| = \begin{cases} C & k = 0 \\ D & 1 \leq k \leq N_{\mathrm{basic}} - 1 \end{cases}, \tag{2.36}$$

then

$$|A(k)| = \begin{cases} 6C & k = 0 \\ 6D & k \in \{6p|0 < p < N/6,\ p \in \mathrm{Z}\} \\ 0 & k \notin \{6p|0 \leq p < N/6,\ p \in \mathrm{Z}\} \end{cases}. \tag{2.37}$$

Given that

$$\left|U_{\mathrm{special}}(k)\right| = \begin{cases} 2\sqrt{3} & k = 1, 5 \\ 0 & k = 0, 2, 3, 4 \end{cases}, \tag{2.38}$$

$$|B(k)| = \begin{cases} 2\sqrt{3}N_{\mathrm{basic}} & k \in \{N_{\mathrm{basic}}, 5N_{\mathrm{basic}}\} \\ 0 & k \notin \{N_{\mathrm{basic}}, 5N_{\mathrm{basic}}\} \end{cases}, \tag{2.39}$$

it can be shown (Tan 2013) that

$$|U(k)| = \begin{cases} 2\sqrt{3}C & k \in \{N_{\mathrm{basic}},\ N - N_{\mathrm{basic}}\} \\ 2\sqrt{3}D & k \in \{1+6p,\ 5+6p|p \in \mathrm{Z}\}, k \notin \{N_{\mathrm{basic}},\ N - N_{\mathrm{basic}}\} \\ 0 & k \notin \{1+6p,\ 5+6p|p \in \mathrm{Z}\} \end{cases}. \tag{2.40}$$

To obtain $u_{\mathrm{basic}}(i)$ which satisfies Eq. 2.36, $u_{\mathrm{basic}}(i)$ can be generated from the classes of MLB, QRB, HAB and TPB signals.

Theorem 2.1 *(Existence of direct synthesis signals) (Tan 2013) A direct synthesis signal* $u_{ds}(i) = u(i)$ *with* $C = 1$, $D = \sqrt{\frac{N_{basic}^2 - 1}{N_{basic} - 1}}$ *and* $N = 6N_{basic}$ *exists for* $N_{basic} \in$

$\{5+6p,\ 7+6p | p \geq 0,\ p \in Z\} \cap (N_{MLB} \cup N_{QRB} \cup N_{HAB} \cup N_{TPB})$, where N_{MLB}, N_{QRB}, N_{HAB} and N_{TPB} are the sets consisting of available periods for MLB, QRB, HAB and TPB signals, respectively, such that $N_{MLB} = \{2^n - 1 | n > 1, n \in Z^+\}$, $N_{QRB} = \{4n - 1 | 4n - 1 \text{ prime}, n \in Z^+\}$, $N_{HAB} = \{4n^2 + 27 | 4n^2 + 27 \text{ prime}, n \in Z^+\}$ and $N_{TPB} = \{n(n+2) | n \text{ prime}, n+2 \text{ prime}\}$. In this case, $u_{basic}(i)$ can be obtained by setting it to the MLB, QRB, HAB or TPB signal.

Proof Based on the properties of MLB, QRB, HAB and TPB signals (Everett 1966), $\sum_{i=1}^{N_{\text{basic}}} u_{\text{basic}}(i) = 1$ and $\sum_{i=1}^{N_{\text{basic}}} u_{\text{basic}}^2(i) = N_{\text{basic}}$. Hence, $C = 1$ can be deduced from the mean value of $u_{\text{basic}}(i)$. Applying Parseval's theorem gives $D = \sqrt{\frac{N_{\text{basic}}^2 - 1}{N_{\text{basic}} - 1}}$. If at least one of the above classes of signals exists for a particular period $N_{\text{basic}} \in \{5+6p,\ 7+6p | p \geq 0,\ p \in Z\}$, $u_{\text{ds}}(i)$ must exist for the period $N = 6N_{\text{basic}}$. □

The direct synthesis signal has the following properties (Tan 2013):

- The signal is ternary, with levels +1, 0 and −1.
- The RMS value is $\sqrt{2/3}$ which is equal to the maximum theoretical limit for the ternary signals with harmonic multiples of two and three suppressed. This gives a PIPS value of 81.65%.
- The *j*th moment, $M_j \equiv E[u^j(i)]$, $j \in Z^+$, is 0 for *j* odd and 2/3 for *j* even.
- The DFT magnitude is given by

$$|U(k)| = \begin{cases} 2\sqrt{3} & k \in \{N_{\text{basic}},\ N - N_{\text{basic}}\} \\ 2\sqrt{\frac{3(N_{\text{basic}}^2 - 1)}{N_{\text{basic}} - 1}} & k \in \{1+6p,\ 5+6p | p \in Z\}, k \notin \{N_{\text{basic}},\ N - N_{\text{basic}}\} \\ 0 & k \notin \{1+6p,\ 5+6p | p \in Z\} \end{cases} \cdot \tag{2.41}$$

- The number of non-suppressed (excited) harmonics with uniform DFT magnitude is given by $\#K = (2N_{\text{basic}} - 2) = (N/3 - 2)$ since from Eq. 2.41, there are two occurrences of magnitude $2\sqrt{3}$ and $(2N_{\text{basic}} - 2)$ occurrences of magnitude $2\sqrt{\frac{3(N_{\text{basic}}^2 - 1)}{N_{\text{basic}} - 1}}$.
- The normalised autocorrelation function, $R_{uu}(n) = \frac{1}{N}\sum_{i=1}^{N} u(i)u(i+n)$, is given by

$$R_{uu}(n) = \begin{cases} 2/3 & n = 0 \\ -2/3 & n = N/2 \\ 1/3 & n = N/6, 5N/6 \\ -1/3 & n = N/3, 2N/3 \\ \alpha(n) & n \notin Nm/6, m \in Z \end{cases}, \tag{2.42}$$

where $\alpha(n)$ is bounded such that $\sup\{|\alpha(n)|\} = 4/N$. The derivation can be found in Tan (2013). The boundedness of the off-peak autocorrelation limits the distortions if the signal is used to estimate a system's impulse response by approximating it to the input–output crosscorrelation function.

An example of the signal $u_{\text{ds}}(i)$ where $u_{\text{basic}}(i)$ is a QRB signal of period $N_{\text{basic}} = 11$ is shown in Fig. 2.10. To generate this signal, a QRB signal of period $N_{\text{basic}} = 11$ is first generated and concatenated 6 times to form $a(i)$ according to Eq. 2.33. The signal $u_{\text{special}}(i) = \begin{bmatrix} 1 & 1 & 0 & -1 & -1 & 0 \end{bmatrix}$ is concatenated 11 times to form $b(i)$ according to Eq. 2.34. Finally, $u_{\text{ds}}(i)$ is formed by taking $a(i) \times b(i)$ according to Eq. 2.32.

In order to further increase the number of available periods, the direct synthesis design is extended to form a class of suboptimal direct synthesis signals (Tan 2013). These signals provide more options in terms of signal period as they are generated by applying QRT signals (Everett 1966) for $u_{\text{basic}}(i)$. QRT signals are more abundant compared with the MLB, QRB, HAB and TPB counterparts. However, the design comes with a drawback of lower signal power (and hence, the signals are suboptimal).

Theorem 2.2 *(Existence of suboptimal direct synthesis signals) (Tan 2013) A suboptimal direct synthesis signal* $u_{sds}(i) = u(i)$ *with* $C = 0$, $D = \sqrt{N_{basic}}$ *and* $N = 6N_{basic}$ *exists for* $N_{basic} \in \{5+6p,\ 7+6p | p \geq 0,\ p \in \mathrm{Z}\} \cap N_{QRT}$*, where* N_{QRT} *is the set consisting of available periods for QRT signals such that* $N_{QRT} = \{4n \pm 1 | 4n \pm 1$ *prime,* $n \in \mathrm{Z}^{+}\}$*. In this case,* $u_{basic}(i)$ *can be obtained by setting it equal to the QRT signal.*

Proof For a QRT signal, $\sum_{i=1}^{N_{\text{basic}}} u_{\text{basic}}(i) = 0$. This gives $C = 0$. Applying Parseval's theorem and noting that $\sum_{i=1}^{N_{\text{basic}}} u_{\text{basic}}^2(i) = N_{\text{basic}} - 1$ lead to $D = \sqrt{N_{\text{basic}}}$. It is obvious that if a QRT signal exists for a particular period $N_{\text{basic}} \in \{5+6p,\ 7+6p | p \geq 0,\ p \in \mathrm{Z}\}$, $u_{\text{sds}}(i)$ must exist for the corresponding $N = 6N_{\text{basic}}$. □

The suboptimal direct synthesis signal has the following properties (Tan 2013):

- The signal is ternary, with levels +1, 0 and −1.
- The RMS value is $\sqrt{\frac{2(N_{\text{basic}}-1)}{3N_{\text{basic}}}} = \sqrt{\frac{2(N-6)}{3N}}$ which is smaller than that of the direct synthesis signal. The derivation is given in Tan (2013). Since suboptimal direct synthesis signals possess lower power compared to direct synthesis signals of the same periods, the latter should be selected in preference over the former, unless the latter is not available for a particular chosen period. Having said that, $\text{rms}(u_{\text{sds}})$ increases rapidly with N. When $N \to \infty$, $\sqrt{\frac{2(N-6)}{3N}} \to \sqrt{\frac{2}{3}}$. For $N \geq 1000$, the difference between $\text{rms}(u_{\text{ds}})$ and $\text{rms}(u_{\text{sds}})$ is negligible (Tan 2013). (At $N = 1000$, $\text{rms}(u_{\text{sds}}) = 0.8140$ which is 99.7% of $\text{rms}(u_{\text{ds}}) = 0.8165$.)
- The jth moment, $\mathrm{M}_j \equiv \mathrm{E}[u^j(i)]$, $j \in \mathrm{Z}^{+}$, is 0 for j odd and $\frac{2(N-6)}{3N}$ for j even.
- The DFT magnitude is given by

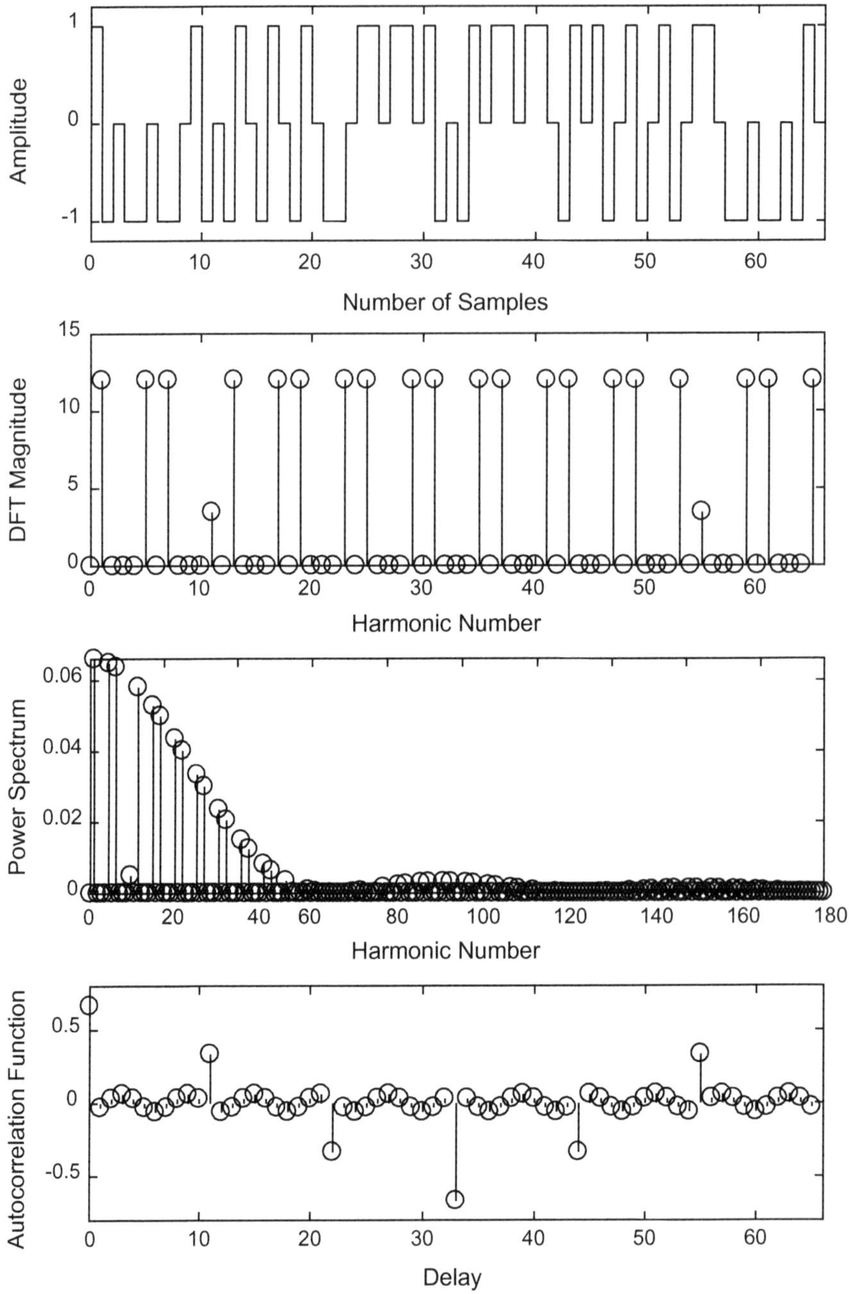

Fig. 2.10 A direct synthesis signal of period $N = 66$

$$|U(k)| = \begin{cases} 0 & k \in \{N_{\text{basic}},\ N - N_{\text{basic}}\} \\ 2\sqrt{3N_{\text{basic}}} & k \in \{1+6p,\ 5+6p|p \in \mathrm{Z}\}, k \notin \{N_{\text{basic}},\ N - N_{\text{basic}}\} \\ 0 & k \notin \{1+6p,\ 5+6p|p \in \mathrm{Z}\} \end{cases}. \tag{2.43}$$

- The number of non-suppressed (excited) harmonics with uniform DFT magnitude is given by $\#K = (2N_{\text{basic}} - 2) = \left(N/3 - 2\right)$ since from Eq. 2.43, there are $(2N_{\text{basic}} - 2)$ occurrences of magnitude $2\sqrt{3N_{\text{basic}}}$.
- The normalised autocorrelation function, $R_{uu}(n) = \frac{1}{N}\sum_{i=1}^{N} u(i)u(i+n)$, is given by

$$R_{uu}(n) = \begin{cases} 2/3 \times \left(\frac{N_{\text{basic}}-1}{N_{\text{basic}}}\right) & n = 0 \\ -2/3 \times \left(\frac{N_{\text{basic}}-1}{N_{\text{basic}}}\right) & n = N/2 \\ 1/3 \times \left(\frac{N_{\text{basic}}-1}{N_{\text{basic}}}\right) & n = N/6, 5N/6 \\ -1/3 \times \left(\frac{N_{\text{basic}}-1}{N_{\text{basic}}}\right) & n = N/3, 2N/3 \\ \beta(n) & n \notin Nm/6, m \in \mathrm{Z} \end{cases} \tag{2.44}$$

where $\beta(n)$ is bounded such that $\sup\{|\beta(n)|\} = 4/N$. The derivation can be found in Tan (2013). The fact that the off-peak autocorrelation is bounded minimises distortions when the signal is used to identify a system by approximating the system's impulse response to the input–output crosscorrelation function.

An example of the signal $u_{\text{sds}}(i)$ where $u_{\text{basic}}(i)$ is a QRT signal of period $N_{\text{basic}} = 11$ is shown in Fig. 2.11. To generate this signal, a QRT signal of period $N_{\text{basic}} = 11$ is first generated and concatenated 6 times to form $a(i)$ according to Eq. 2.33. The signal $u_{\text{special}}(i) = \left[\ 1\ 1\ 0\ -1\ -1\ 0\ \right]$ is concatenated 11 times to form $b(i)$ according to Eq. 2.34. Finally, $u_{\text{sds}}(i)$ is formed by taking $a(i) \times b(i)$ according to Eq. 2.32. Note the similarity of $u_{\text{sds}}(i)$ with the signal $u_{\text{ds}}(i)$ in Fig. 2.10. This similarity is due to the fact that the QRB and the QRT signals differ only by one bit in a period of N_{basic}.

For the direct synthesis method (including its suboptimal version), no knowledge of primitive polynomials is necessary for generating $u_{\text{basic}}(i)$ based on QRB, HAB, TPB or QRT signals. This presents a significant advantage as the signal generation does not rely on the availability of primitive polynomial tables. Thus, there is practically no upper limit to the periods available for signals generated via the direct synthesis approach. Coupled with the advantages of high signal power and short computational time as no exhaustive search or computer optimisation procedure is involved, the direct synthesis approach is highly favourable when ternary perturbation signals with harmonic multiples of two and three suppressed are required.

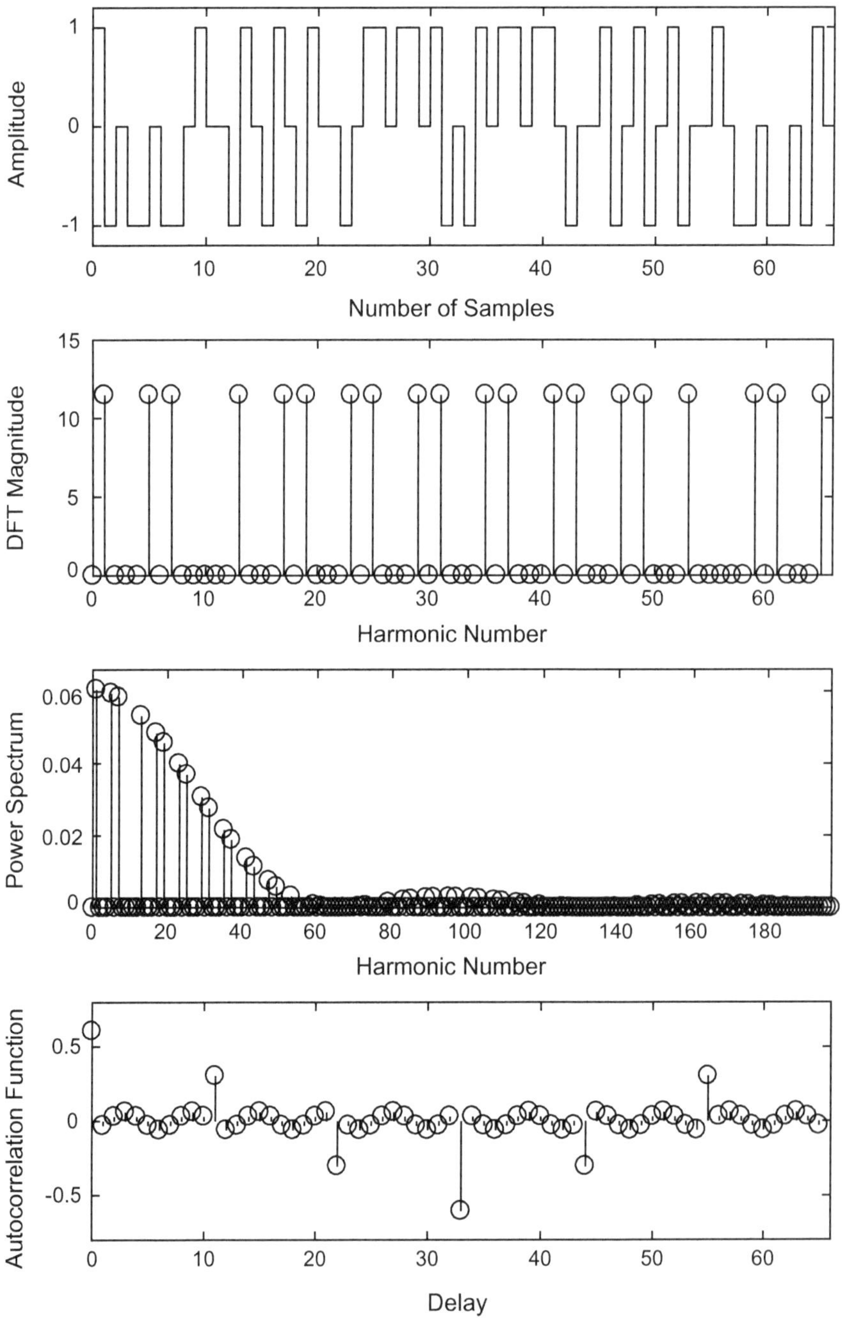

Fig. 2.11 A suboptimal direct synthesis signal of period $N = 66$

Table 2.9 Properties of signals used in the simulation example

Signal	N	rms(u)	#K
S1	1848	0.610	616
S2	1842	0.816	612
S3	1842	0.815	612

2.5 Application Example

A simulation example was carried out using the transfer function $G(z^{-1})$ given by

$$G(z^{-1}) = \frac{-5.667 \times 10^{-5} + 1.566 \times 10^{-4} z^{-1} - 1.155 \times 10^{-4} z^{-2}}{1 - 2.6705 z^{-1} + 2.3730 z^{-2} - 0.7024 z^{-3}} \tag{2.45}$$

under a sampling interval t_s of 0.2 s, as described in Tan (2013). An output nonlinearity of $y_{\text{nonlinear}} = y_{\text{linear}} + 0.2 y_{\text{linear}}^2 + 0.1 y_{\text{linear}}^3$ was added to gauge the performance of the signals for linear identification in the presence of nonlinear distortion. The measurement time T_N was chosen to satisfy $T_N \geq 5 \times$ estimated largest time constant $\approx 5 \times 60$ s = 300 s giving a suitable range for $N = T_N / t_s$ of $1500 \leq N \leq 2000$.

In order to eliminate the effects of even-order nonlinearity and minimise the effects of odd-order nonlinearity, signals with harmonic multiples of two and three suppressed were desired. The classes of pseudorandom signals which are able to meet this frequency-domain specification are PRML and truncated PRML signals (Sect. 2.3) as well as direct synthesis ternary and suboptimal direct synthesis ternary signals (Sect. 2.4). Taking into account the required range for N, three signals formed from different classes were used in the comparison:

- S1 (PRML with $q = 43$, $n = 2$),
- S2 (direct synthesis from a QRB signal with $N_{\text{basic}} = 307$) and
- S3 (suboptimal direct synthesis from a QRT signal with $N_{\text{basic}} = 307$).

For the truncated PRML signals, the closest periods are 1098 and 2286 which fall outside the range $1500 \leq N \leq 2000$. These were not utilised as too short a period would lead to low-frequency resolution, whereas too long a period would unnecessarily lengthen experimentation time.

Properties of the signals S1, S2 and S3 are summarised in Table 2.9. The signals were scaled to amplitudes of ± 1 and share similar properties except that S1 has a considerably lower RMS value compared with S2 and S3.

The transfer functions were identified via maximum likelihood estimation (Pintelon and Schoukens 2012) based on a single period of steady-state data. Data beyond 2 Hz were discarded due to the low SNR at high frequency. The actual and estimated FRFs are illustrated in Fig. 2.12 for simulations using output additive white Gaussian noise of rms(noise) = 1×10^{-4} and 5×10^{-4}. The graphs for S3 were not included as they were very similar to those obtained for S2. When rms(noise) = 1×10^{-4}, the esti-

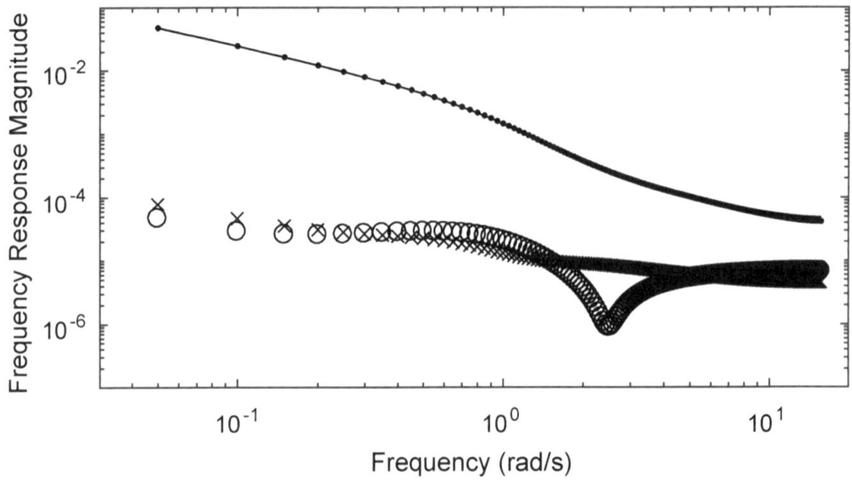

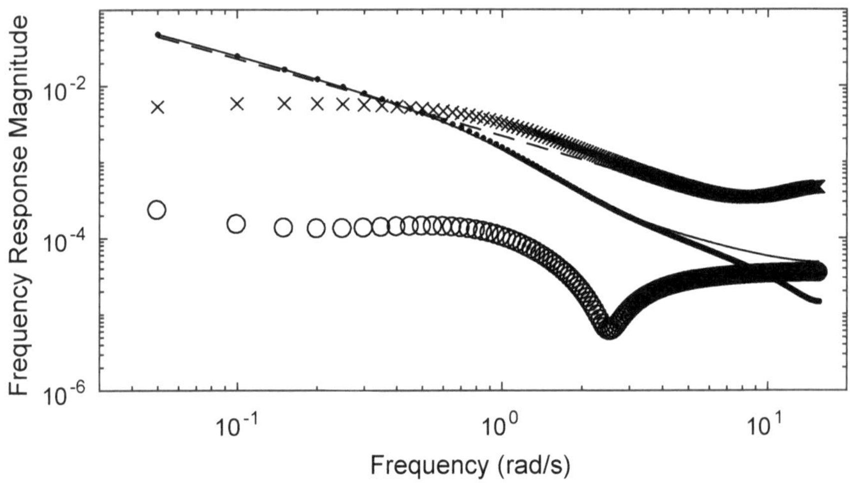

Fig. 2.12 FRFs of the simulated system for rms(noise) = 1×10^{-4} (top) and rms(noise) = 5×10^{-4} (bottom). Solid line: actual; dashed line: estimates using S1; dotted line: estimates using S2; crosses: magnitude of complex error in estimates using S1; circles: magnitude of complex error in estimates using S2 (Reproduced with permission from Tan © 2013 Elsevier)

mates using S1 and S2 were of comparable quality. However, when rms(noise) = 5×10^{-4}, S2 resulted in greater estimation accuracy due to the higher RMS value of the signal resulting in a larger SNR. In particular, the effects of noise were very significant at higher frequencies when S1 was applied. The estimated transfer function was unstable, and the corresponding magnitude of the complex error was significantly larger than that obtained using S2.

The value of the direct synthesis approach is evident from this example, both in terms of providing a flexible choice of signal period and in generating a signal with high RMS value within amplitude constraints for greater robustness in the presence of noise and nonlinear distortion.

References

Barker HA (1993) Design of multi-level pseudo-random signals for system identification. In: Godfrey K (ed) Perturbation signals for system identification. Prentice Hall, Englewood Cliffs, NJ

Barker HA, Zhuang M (1997) Design of pseudo-random perturbation signals for frequency-domain identification of nonlinear systems. IFAC Proc Volumes 30:1649–1654

Charters P (2009) Generalizing binary quadratic residue codes to higher power residues over larger fields. Finite Fields Appl 15:404–413

Dai Z, Gong G, Song H-Y (2009) A trace representation of binary Jacobi sequences. Discrete Math 309:1517–1527

Davidson JN, Stone DA, Foster MP (2015) Real-time prediction of power electronic device temperatures using PRBS-generated frequency-domain thermal cross coupling characteristics. IEEE Trans Power Electron 30:2950–2961

Debenjak A, Boškoski P, Musizza B, Petrovčič J, Juričić D (2014) Fast measurement of proton exchange membrane fuel cell impedance based on pseudo-random binary sequence perturbation signals and continuous wavelet transform. J Power Sources 254:112–118

Egidi L, Manzini G (2013) Better spaced seeds using quadratic residues. J Comput Syst Sci 79:1144–1155

Everett D (1966) Periodic digital sequences with pseudonoise properties. G.E.C. Journal 33:115–126

Godfrey K (1993) Introduction to perturbation signals for time-domain system identification. In: Godfrey K (ed) Perturbation signals for system identification. Prentice Hall, Englewood Cliffs, NJ

Godfrey KR, Tan AH, Barker HA, Chong B (2005) A survey of readily accessible perturbation signals for system identification in the frequency domain. Control Eng Pract 13:1391–1402

Golomb SW (2017) Shift register sequences: secure and limited-access code generators, efficiency code generators, prescribed property generators, mathematical models. World Sci, Singapore

Lee C-D, Huang Y-P, Chang Y, Chang H-H (2015) Perfect Gaussian integer sequences of odd period 2^m–1. IEEE Signal Process Lett 22:881–885

Neshvad S, Chatzinotas S, Sachau J (2015) Wideband identification of power network parameters using pseudo-random binary sequences on power inverters. IEEE Trans Smart Grid 6:2293–2301

Nguyen STN, Gong J, Lambert MF, Zecchin AC, Simpson AR (2018) Least squares deconvolution for leak detection with a pseudo random binary sequence excitation. Mech Syst Signal Process 99:846–858

Pintelon R, Schoukens J (2012) System identification: a frequency domain approach. Wiley, Hoboken, NJ

Roinila T, Yu X, Verho J, Li T, Kallio P, Vilkko M, Gao A, Wang Y (2014) Methods for rapid frequency-domain characterization of leakage currents in silicon nanowire-based field-effect transistors. Beilstein J Nanotechnol 5:964–972

Tan AH (2007) Design of truncated maximum length ternary signals where their squared versions have uniform even harmonics. IEEE Trans Autom Control 52:957–961

Tan AH (2013) Direct synthesis of pseudo-random ternary perturbation signals with harmonic multiples of two and three suppressed. Automatica 49:2975–2981

Tan AH, Foo MFL (2006) Ternary pseudorandom signal design for uniform excitation and reduced effect of nonlinear distortion. Electron Lett 42:676–677

Tan AH, Godfrey KR (2002) The generation of binary and near-binary pseudorandom signals: an overview. IEEE Trans Instrum Meas 51:583–588

Tan AH, Godfrey KR, Barker HA (2005) Design of computer-optimized pseudo-random maximum length signals for linear identification in the presence of nonlinear distortions. IEEE Trans Instrum Meas 54:2513–2519

Chapter 3
Design of Computer-Optimised Signals for Linear System Identification

3.1 Multisine Sum of Harmonics Signals

Multisine signals can be written mathematically as

$$u(t) = \sum_{p \in B} A_p \cos\left(2\pi f_p t + \phi_p\right), 0 \leq t \leq T_N, \tag{3.1}$$

where B denotes the set consisting of nonzero harmonics, while A_p, f_p and ϕ_p represent the amplitude, frequency and phase associated with harmonic p. T_N is the period of the signal. Multisine signals can be generated either in the time domain by implementing Eq. 3.1 directly, or in the frequency domain by specifying the amplitudes and phases of each harmonic and then taking the inverse DFT. For a signal comprising many harmonics, the latter is likely to be more computationally efficient. A multisine signal can take any value within the range between its minimum and maximum values hence giving it complete flexibility in terms of frequency domain specifications. This has led to many applications using multisine signals, such as the modelling of Li-ion battery equivalent circuit (Widanage et al. 2016), the identification of hydro-static drive-line (Stoev and Schoukens 2016), the identification of an X-ray system (van der Maas et al. 2016), and the identification of bioimpedance by means of electrical impedance spectroscopy (Sanchez et al. 2013). An application on a mist reactor system (Cham et al. 2017) is described in Sect. 5.4.2.

Several forms of multisine exist. For the generation of a continuous-time signal from sampled data, there are two popular assumptions. The first is band-limited assumption, such that the time domain signal will be generated from sampled data with very high clock frequency or with a reconstruction filter assuming exact samples of a band-limited signal. Thus, for the final continuous-time signal, the values of the signal in between the calculated samples may be larger than the neighbouring samples. The second is ZOH assumption, that the continuous-time signal is piecewise constant, having passed through a ZOH, which has a $(\sin^2 x)/x^2$ power spectrum

A. H. Tan and K. R. Godfrey, *Industrial Process Identification*, Advances in Industrial Control, https://doi.org/10.1007/978-3-030-03661-4_3

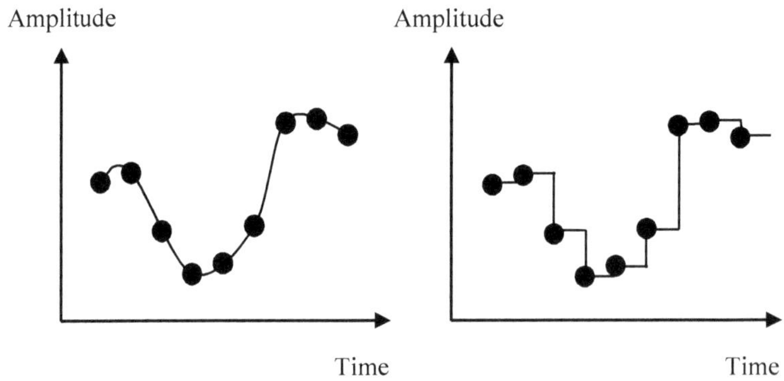

Fig. 3.1 Intersampling behaviour with band-limited assumption (left) and ZOH assumption (right). The dots represent the sampled data

envelope. This is the normal assumption for applications involving discrete controller design. Pre-compensation for the shape of the ZOH spectrum is necessary. For discrete-time signal, there is no need to deal with the intersampling behaviour and this is the simplest case for the multisine design. ZOH pre-compensation is not necessary, as in the case of band-limited assumption. Figure 3.1 illustrates the intersampling behaviour with band-limited assumption and ZOH assumption.

To design a multisine signal, the set of harmonics in which the signal power is required is specified, together with the desired amplitude spectrum, including whether ZOH pre-compensation is required. For linear system identification, the objective of the optimisation is to adjust the relative phases of the harmonics in order to minimise the peak-to-peak amplitude of the signal, thus minimising the crest factor and maximising PIPS. In some literature particularly in the area of communications, this is referred to as the peak-to-average power ratio (PAPR) reduction problem, where PAPR is simply the square of the crest factor.

Several methods are available in the literature relating to the design of the relative phases of the harmonics. The worst choice of phasing is setting all the phases of the harmonics to zero such that $\phi_p = 0$ for all values of p since the peaks of all the excited harmonics coincide at $t = 0$. This gives very poor crest factor of $\sqrt{2F}$, where F is the number of specified nonzero harmonics. Random-phase design is obtained by setting the phases randomly according to a uniform distribution between 0 and 2π. This design yields crest factors on the order of $\sqrt{\log F}$ (Gersho et al. 1979).

Deterministic formulas for the design of relative phases were proposed in the literature since several decades ago. In the Shapiro–Rudin phase design, the phases can take only two values and are calculated using (Shapiro 1951; Rudin 1959)

$$\phi_k = \begin{cases} 0 & r_k = 1 \\ \pi & r_k = -1 \end{cases}, \tag{3.2}$$

where r_k is generated by the process of concatenation and inversion. The initial string is $[r_1 \quad r_2] = [1 \quad 1]$. This is concatenated with the same sequence where the second half is inverted. For example, for a signal with four excited harmonics, $[r_1 \quad r_2 \quad r_3 \quad r_4] = [1 \quad 1 \quad 1 \quad -1]$; for a signal with eight excited harmonics, $[r_1 \quad r_2 \quad r_3 \quad r_4 \quad r_5 \quad r_6 \quad r_7 \quad r_8] = [1 \quad 1 \quad 1 \quad -1 \quad 1 \quad 1 \quad -1 \quad 1]$ and so on. If the number of harmonics is a power of 2, it has been proven mathematically by Boyd (1986) that the crest factor ≤ 2. It should be noted that the relationship in Eq. 3.2 applies to the case where the harmonics are equally spaced and with equal amplitude. Although the formula allows straightforward extension to cases where this is not true, the results may not be as good.

In the Newman phase design (Newman 1965), the phases are calculated using

$$\phi_k = \frac{(k-1)^2\pi}{F}. \tag{3.3}$$

It was reported by Boyd (1986) that numerical investigations indicate that Newman phase design generally yields smaller crest factors than the Shapiro–Rudin phase design. However, the study was carried out for lowpass and bandpass spectra where all harmonics are present within the frequency band of interest, and the amplitudes of all the harmonics are equal. This is the specification for which the formula Eq. 3.3 was proposed.

The Schroeder phase design (Schroeder 1970) uses

$$\phi_k = \phi_1 - 2\pi \sum_{l=1}^{k-1} (k-l) p_l, \tag{3.4}$$

where ϕ_1 can be chosen arbitrarily and p_l is the relative power of the lth harmonic scaled such that $\sum_{k=1}^{F} p_k = 1$. To obtain the phases, without loss of generality, we can set the initial phase $\phi_1 = 0$. For example, for a specification with $F = 4$, the phases are calculated using $\phi_2 = \phi_1 - 2\pi \sum_{l=1}^{1} (2-l) p_l = -2\pi p_1$, $\phi_3 = \phi_1 - 2\pi \sum_{l=1}^{2} (3-l) p_l = -2\pi(2p_1 + p_2)$ and $\phi_4 = \phi_1 - 2\pi \sum_{l=1}^{3} (4-l) p_l = -2\pi(3p_1 + 2p_2 + p_3)$. For the case of uniform spectra with all the excited harmonics having the same power, $p_k = 1/F$. Thus

$$\phi_k = \phi_1 - \frac{2\pi}{F} \sum_{l=1}^{k-1} (k-l). \tag{3.5}$$

Using the simplification that $1 + 2 + 3 + \cdots + k - 1 = \frac{k(k-1)}{2}$,

$$\phi_k = \phi_1 - \frac{k(k-1)\pi}{F}. \tag{3.6}$$

It can be seen that there is some similarity with the Newman phasing. In general, Schroeder phase design gives low crest factors provided the amplitude spectrum is uniform and the harmonics are either consecutive or consecutive odd.

The deterministic techniques are generally limited to specifications with equal amplitude and equal harmonic spacing. To overcome these constraints, phase design based on computer-optimisation was proposed. The algorithm used in `msinclip` in the Frequency Domain System Identification Toolbox (Kollár 1994) is based on swapping between the time domain and the frequency domain (Van der Ouderaa et al. 1988). To start the time-frequency swapping algorithm, the user can specify either Schroeder phases or random phases. External input of initial phases is also possible, if necessary. The algorithm starts with predefined amplitude spectrum, and an initial set of phases. These are combined in the frequency domain and then inverse DFT is performed. The time domain signal is clipped at a suitable level and DFT is performed. Then in the frequency domain, the new phases are retained while the amplitude spectrum is reset to the predefined values. The procedure is iterated until there is no further decrease in the crest factor. The algorithm allows 'snowing' to be incorporated. 'Snowing' refers to the addition of power outside the frequency band of interest, in order to decrease the crest factor. However, it has a downside that some power is lost to frequencies outside the band of interest.

The infinity norm algorithm proposed by Guillaume et al. (1991) is based on Polya's algorithm and attempts to minimise the l_p-norm given by

$$l_p(u) = \left[\frac{1}{T_N} \int_0^{T_N} |u(t)|^p \mathrm{d}t \right]^{1/p}, p \geq 1. \tag{3.7}$$

Generally, $p = 2$ at the start of the procedure. Once a set of phases has been obtained which minimises Eq. 3.7 for $p = 2$, the value of p is increased to 4. When Eq. 3.7 has been minimised for $p = 4$, the value of p is increased to 8. Throughout the procedure, p takes the values 2, 4, 8, 16, 32, …. This algorithm is available through the `crestmin` function in the Frequency Domain System Identification Toolbox (Kollár 1994). The algorithm can accommodate 'snowing'. According to Pintelon and Schoukens (2012), the infinity norm algorithm performs better than the time-frequency swapping algorithm.

In terms of frequency domain specification, a full multisine has all harmonics being excited within the bandwidth of interest. An odd multisine is an inverse-repeat signal with even harmonics suppressed. It can be generated if the signal period is set to an even number. This property ensures that the even harmonics of the signal are suppressed, so that odd order and even order nonlinear distortion in the system output may be separated, and the effect of the latter on the estimate of the linear dynamics can be completely eliminated. Specification with harmonic multiples of two and three suppressed is also common, with the additional benefit of reducing the effects of odd order nonlinear distortion on the linear estimate. Besides these, several other designs are available. An example is the odd-odd design (Schoukens et al. 2001) where the excited harmonics are set at $4p + 1$ (these harmonics are 1, 5, 9, 13, 17, 21, …). At lines $4p + 1$, the output consists of the linear contribution plus odd order nonlinear distortions; at lines $4p + 2$ and $4p + 4$: only the even order nonlinear distortions appear; at lines $4p + 3$: only the odd order nonlinear distortions appear. Hence, it is possible to detect and separate the effects of odd order and even order nonlinearities. The odd-odd multisine was recently applied for characterising a signal generator for the testing of medium-voltage measurement transducers (Faifer et al. 2015).

The sparse odd design is a design where some of the odd harmonics are suppressed, in addition to the even harmonics. Strictly speaking, both the design with harmonic multiples of two and three suppressed and the odd-odd design are also sparse odd designs, but the term generally refers to cases where there is no clear pattern in the suppression of the odd harmonics. For example, Kulesza (2014) utilised multisines with harmonics 1, 5, 9, and 13; and 1, 3, 7, 11 and 15 in the study of the dynamic behaviour of cracked rotor. In another application, Oliva Uribe et al. (2018) selected the harmonics 1, 5, 7, 9, 11, 15, 19, 21, 23, … in the multisine used to drive a piezoelectric bimorph sensor. In the design, for every four consecutive odd frequency lines, one harmonic was randomly selected for suppression.

The performance of the Schroeder phase design, the time-frequency swapping algorithm implemented in `msinclip` and the infinity norm algorithm implemented in `crestmin` is compared for four specifications:

Specification A: 100 consecutive harmonics (1, 2, 3, 4, …, 100), $N = 250$;
Specification B: 100 consecutive odd harmonics (1, 3, 5, 7, …, 199), $N = 500$;
Specification C: 100 consecutive harmonics which are not multiples of two and three (1, 5, 7, 11, …, 299), $N = 750$;
Specification D: 10 logarithmically spaced harmonics (1, 2, 4, 8, …, 512), $N = 1280$.

The specifications all have equal DFT magnitude for the excited harmonics. The highest specified harmonic is set to approximately $0.4\,N$, which is close to the highest usable harmonic in a spectrum analyser that uses an analog anti-aliasing filter.

As an example, to obtain a multisine with Schroeder phase design for Specification A, the following codes can be used in MATLAB®. (MATLAB® is a registered product of The MathWorks, Inc.)

```
%set the frequency vector f and period N
f=[1:100];N=250;
%calculate the required amplitude with ZOH
%pre-compensation
%if ZOH pre-compensation is not required,
%set amp=ones(1,length(f));
for k=1:length(f)
   amp(k)=pi*f(k)/N/(sin(pi*f(k)/N));
end

%calculate Schroeder phases
pk=abs(amp.^2)/sum(abs(amp.^2));
phi(1)=0;
for k=2:length(f)
```

```
    phi(k)=-2*pi*[k-1:-1:1]*[pk(1:k-1)]';
end

%obtain the signal u through the frequency domain
U=zeros(N,1);
U(f+1)=amp.*exp(j.*phi);
U(N-f+1)=conj(U(f+1));
u=real(ifft(U));
%scale to obtain RMS of 1
u=u/sqrt(mean(u.^2));

%For Specification B, the frequency vector and period need to
%be set to
f=[1:2:199];N=500;

%For Specification C, the frequency vector and period need to
%be set to
f=[1:2:299];f(2:3:150)=[];N=750;

%For Specification D, the frequency vector and period need to
%be set to
f=2.^[0:9];N=1280;
```

For time-frequency swapping and infinity norm algorithms, refer to Sect. 8.3.1. Both algorithms were run at their default number of iterations, that is, 200 for the time-frequency swapping algorithm and 50 for each p value for the infinity norm algorithm. The stopping value of p is 256 by default. The starting phases were set to those given by the random-phase design, which is the default setting.

The performance measures obtained both with and without ZOH pre-compensation are shown in Tables 3.1 and 3.2, respectively. Note that for the latter case (suitable for band-limited and discrete designs), PIPSE and EMINE are not applicable as these measures are defined assuming ZOH pre-compensation (see Sect. 1.4). The DFT magnitude of Specification A is plotted in Fig. 3.2. The multisine signals are shown in Figs. 3.3 and 3.4 for the cases with and without ZOH pre-compensation, respectively. The signals were all scaled to have an RMS value of 1. The DFT magnitude of Specification D is plotted in Fig. 3.5. The multisine signals are shown in Figs. 3.6 and 3.7 for the cases with and without ZOH pre-compensation, respectively. The logarithmic spacing (including quasi-logarithmic spacing) is useful, for example, in the study of music (Kazazis et al. 2017), in bioimpedance spectroscopy (Sanchez et al. 2013; Yang et al. 2015) and for applications where a large frequency range needs to be covered (Pintelon et al. 2014).

From Tables 3.1 and 3.2, it can be seen that the infinity norm algorithm is able to give the lowest crest factor values, followed by the time-frequency swapping method. The low crest factor of the infinity norm algorithm leads to relatively high PIPS and PIPSE values. The Schroeder phase design results in the highest crest factor values. Its effectiveness is limited especially for the logarithmically spaced specification. However, it is worth highlighting that this method is useful in applications where optimisation time is a constraint as well as in setting the starting phases for further optimisation. Higher crest factors values are observed for Specification D compared to Specifications A, B and C as the spacing of the harmonics is not linear and there are fewer excited harmonics thus giving lower flexibility in the optimisation of the relative phases. The value of EMINE is 100% in all cases as the multisine signal is able to meet all the frequency domain specifications exactly. The time domain plots in Figs. 3.3, 3.4, 3.6 and 3.7 show that the peak values are largest for the Schroeder phase design, followed by the time-frequency swapping algorithm and finally, the infinity norm algorithm.

It is important to point out that by minimising the crest factor, the multisine no longer has a Gaussian amplitude distribution. In applications involving nonlinear systems, it may be useful to have a Gaussian amplitude distribution, in which case random-phase multisines should be applied (see Sect. 5.2 on the identification of the best linear approximation). However, it is possible to have a trade-off between the two requirements by specifying a small number of iterations in the time-frequency swapping and infinity norm algorithms. For the infinity norm algorithm, the maximum value of p should be set small as well if this trade-off is to be achieved. The effects of the number of iterations and the maximum value of p on the amplitude distribution can be observed from Fig. 3.8. An increase in the number of iterations leads to a distribution which deviates further from a Gaussian distribution as the optimisation algorithm aims to minimise the crest factor resulting in a larger portion

Table 3.1 Performance measures of multisine signals generated using different phase designs with ZOH pre-compensation

Specification	Phase design	Crest factor	PIPS (%)	PIPSE (%)	EMINE (%)
A	Schroeder	1.69	61.00	55.24	100
A	Time-frequency swapping	1.41	71.50	64.75	100
A	Infinity norm	1.18	84.99	76.95	100
B	Schroeder	1.67	59.87	54.30	100
B	Time-frequency swapping	1.35	74.11	67.21	100
B	Infinity norm	1.20	83.59	75.80	100
C	Schroeder	2.16	46.33	42.02	100
C	Time-frequency swapping	1.60	62.32	56.52	100
C	Infinity norm	1.36	73.79	66.92	100
D	Schroeder	3.24	32.02	30.63	100
D	Time-frequency swapping	2.48	42.42	40.57	100
D	Infinity norm	2.25	44.46	42.52	100

Table 3.2 Performance measures of multisine signals generated using different phase designs without ZOH pre-compensation

Specification	Phase design	Crest factor	PIPS (%)
A	Schroeder	1.60	63.36
A	Time-frequency swapping	1.57	64.47
A	Infinity norm	1.18	85.13
B	Schroeder	1.66	60.26
B	Time-frequency swapping	1.51	66.08
B	Infinity norm	1.19	84.22
C	Schroeder	2.03	49.23
C	Time-frequency swapping	1.80	55.63
C	Infinity norm	1.35	73.85
D	Schroeder	3.31	32.10
D	Time-frequency swapping	2.65	39.20
D	Infinity norm	2.25	44.49

of the signal being close to the maximum and minimum amplitudes. A similar effect is observed when the maximum value of p is increased.

3.2 Discrete Interval Binary and Discrete Interval Ternary Signals

DIB and DIT signals are computer-optimised signals with a specific number of signal levels (two levels and three levels, respectively). The objective of the optimisation is to force as much power as possible into the specified harmonics. The function is based on an algorithm of van den Bos and Krol (1979). It is very similar to the time-frequency swapping algorithm. The algorithm starts with predefined amplitude spectrum, and an initial set of phases. These are combined in the frequency domain and then inverse DFT is performed. For the DIB signal, the time domain signal is hard-limited into two levels such that all bits with positive values are set to $+1$ and all bits with negative values are set to -1. Those with zero value can be set to either $+1$ or -1. For the DIT signal, quantisation is performed to obtain a ternary signal (refer to Sect. 3.3 for more details on quantisation). The DFT is performed on the time domain signal. In the frequency domain, the new phases are retained while the amplitude spectrum is reset to the predefined values. A recent work applying the DIB signal for online grid impedance measurement is described by Roinila et al. (2014).

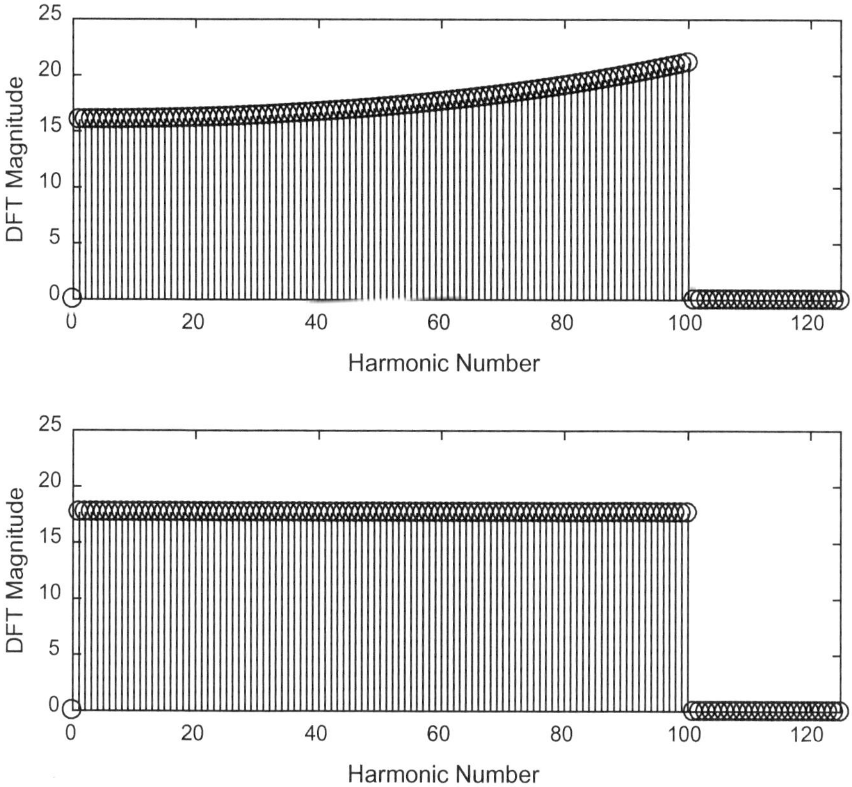

Fig. 3.2 Specification A with ZOH pre-compensation (top) and without ZOH pre-compensation (bottom)

The performance measures for Specifications A to D (see Sect. 3.1) are summarised in Tables 3.3 and 3.4. The computations were carried out using the `dibs` and `dits` functions in the Frequency Domain System Identification Toolbox (Kollár 1994). The default settings were applied, namely 25 random phases as starting points with no limit on the maximum number of iterations required for convergence for each set of starting phases.

In Tables 3.3 and 3.4, P_{uf} is the useful power as a fraction of the total power, which is an output of the `dibs` and `dits` functions. In the case without ZOH pre-compensation, the power is not shaped by the $(\sin^2 x)/x^2$ power spectrum envelope. The signals were all scaled to have an RMS value of 1. For details of the usage of software to generate these signals, please refer to Sect. 8.3.2. Note that for Specification C, DIB signals could not be applied since a binary signal cannot satisfy the requirement of having harmonic multiples of two and three suppressed.

From Tables 3.3 and 3.4, the value of PIPS for a DIB signal is close to 100%, because the signal is binary and the mean of the signal is either zero or very close to

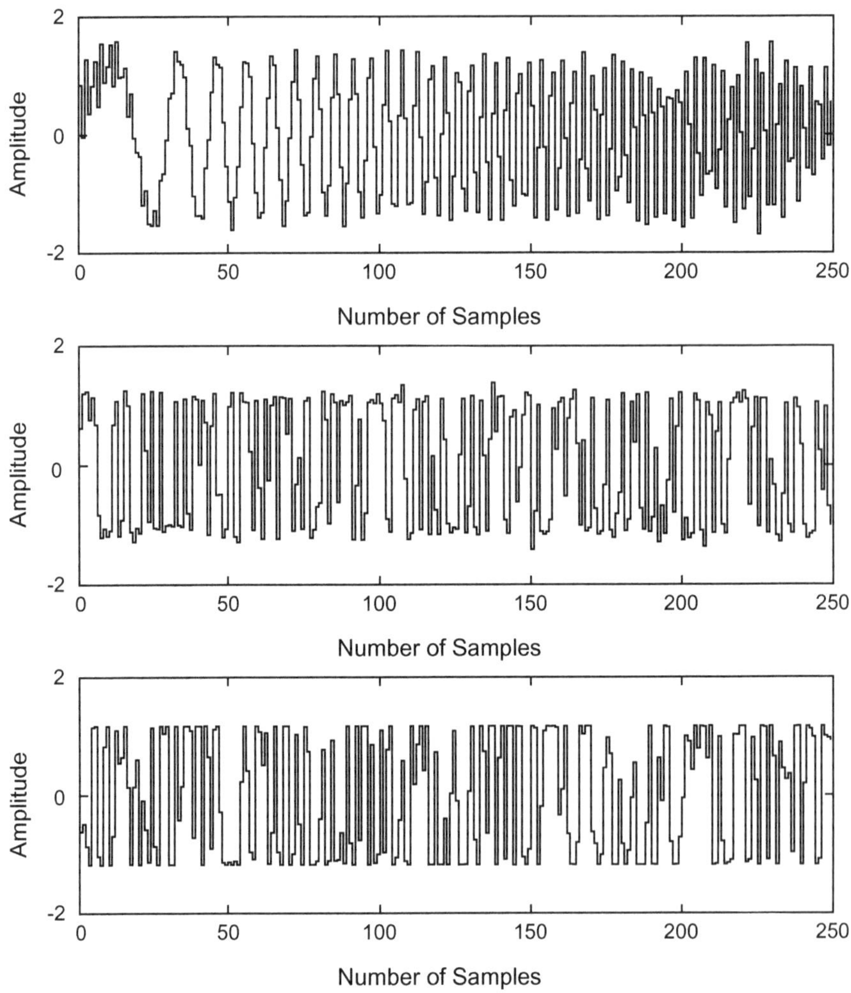

Fig. 3.3 Multisine signals for Specification A with ZOH pre-compensation. Top: Schroeder; middle: time-frequency swapping; bottom: infinity norm

it. Interestingly, the `dits` function can occasionally result in a binary signal. This was the case for Specification D. Due to the constraint on the number of signal levels, the amplitude spectrum will not match the specifications exactly. Thus, the value of PIPSE is less than 100% for any harmonic specification. The value of EMINE can also be low for some harmonic specifications. Using a DIT signal may improve the value of EMINE. This can be observed in Figs. 3.9 and 3.10 where the power spectra obtained using the ternary signal have smaller fluctuations in the specified nonzero harmonics. However, this is often at the expense of lower PIPS, PIPSE and P_{uf}.

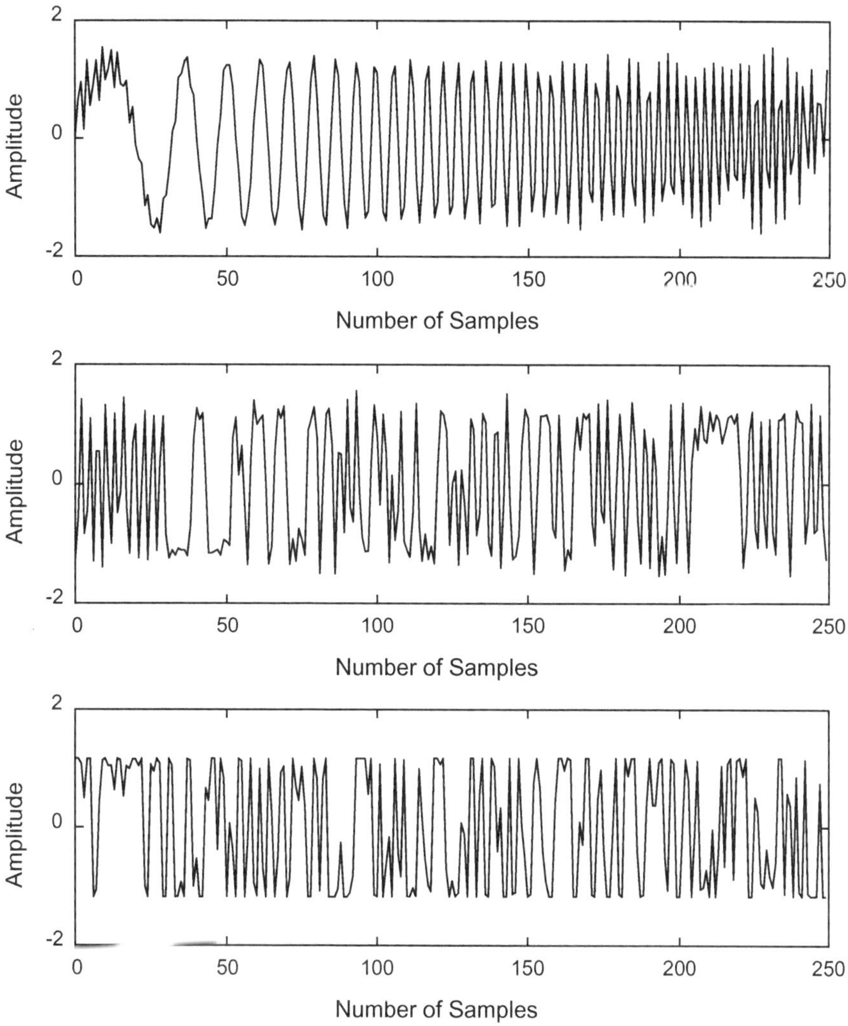

Fig. 3.4 Multisine signals for Specification A without ZOH pre-compensation. Top: Schroeder; middle: time-frequency swapping; bottom: infinity norm

3.3 Multilevel Multiharmonic Signals

DIB and DIT signals have clearly defined peak-to-peak amplitude, but a significant amount of the total power appears in non-specified harmonics. In contrast, multisine signals have more power in the specified harmonics (100% of the power if 'snowing' is not applied), but they also have higher peak-to-peak amplitudes for a given harmonic specification. MLMH signals are designed to retain the advantages of each

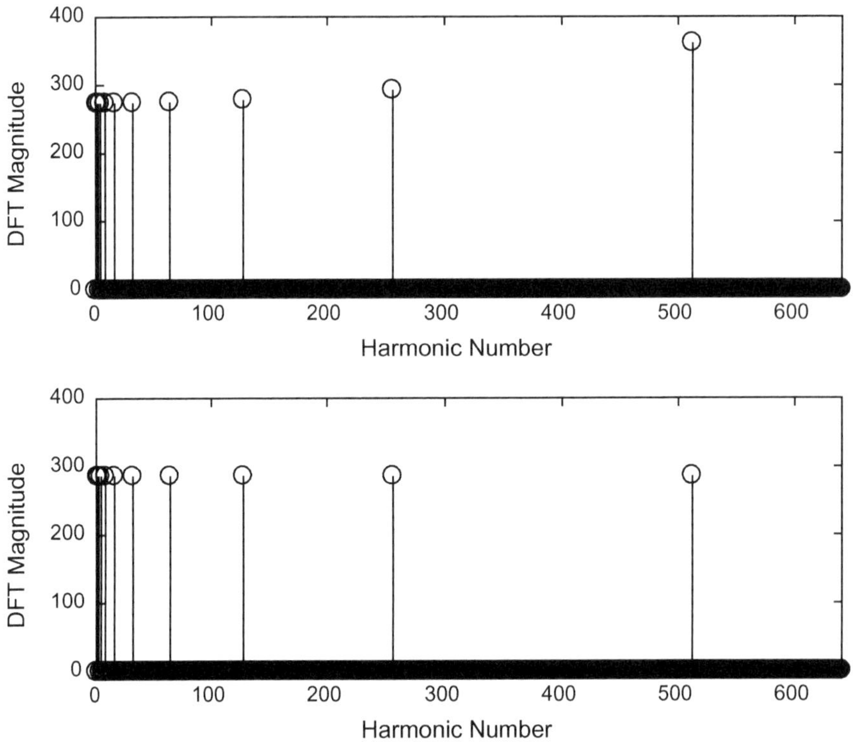

Fig. 3.5 Specification D with ZOH pre-compensation (top) and without ZOH pre-compensation (bottom)

Table 3.3 Performance measures of DIB and DIT signals generated with ZOH pre-compensation

Specification	Signal type	Crest factor	PIPS (%)	PIPSE (%)	EMINE (%)	P_{uf} (%)
A	DIB	1	100	88.44	39.95	78.22
A	DIT	1.03	96.91	86.65	54.97	75.08
B	DIB	1	100	88.64	41.23	78.57
B	DIT	1.15	86.95	78.09	58.02	60.98
C	DIB	N/A	N/A	N/A	N/A	N/A
C	DIT	1.26	77.33	71.39	54.18	50.97
D	DIB	1	99.95	78.48	97.74	61.59
D	DIT	1	99.98	78.15	98.69	61.07

type of signal, while reducing the disadvantages (McCormack et al. 1995; Godfrey et al 2005).

MLMH signals can be generated using the `multilev_new` algorithm (see Sect. 8.4) which aims to achieve a compromise between maximising PIPSE and EMINE

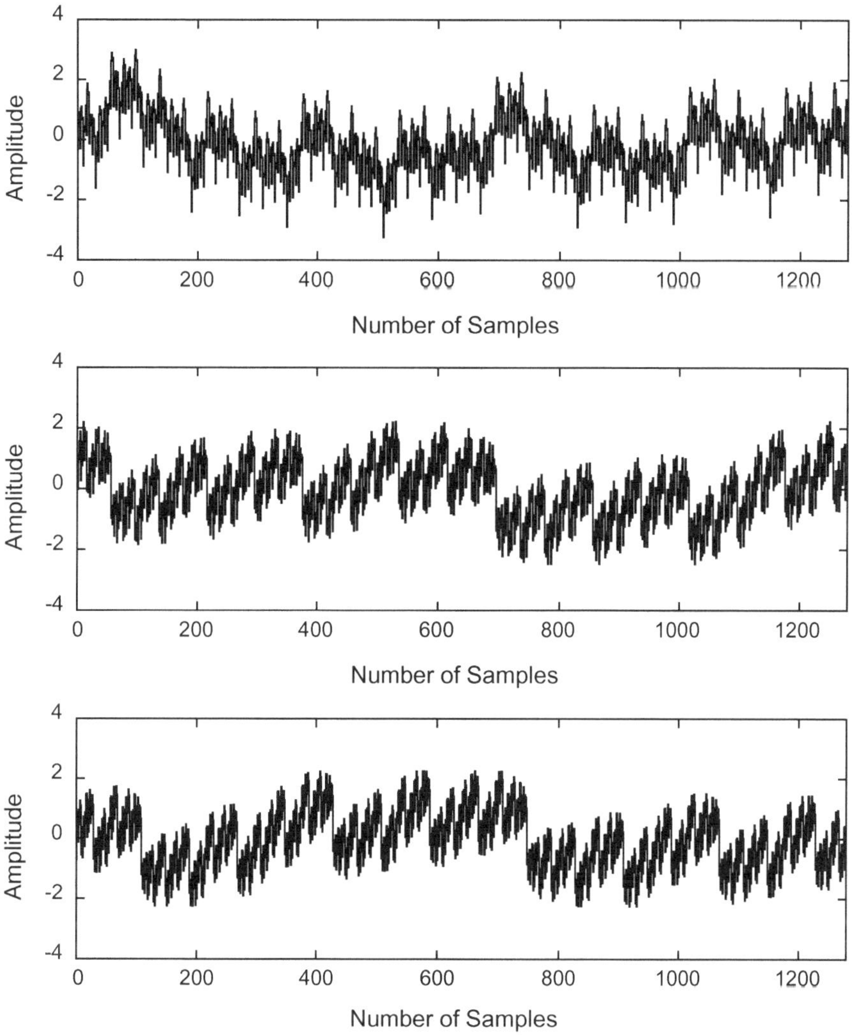

Fig. 3.6 Multisine signals for Specification D with ZOH pre-compensation. Top: Schroeder; middle: time-frequency swapping; bottom: infinity norm

(Tan and Godfrey 2004). PIPSE and EMINE are given equal weight in the algorithm. In the case without ZOH pre-compensation, similar measures to these are applied, but without the $(\sin^2 x)/x^2$ power spectrum envelope. The same time-frequency swapping algorithm used for the design of multisine, DIB and DIT signals is employed. However, the number of signal levels M is an input parameter.

Quantisation plays an important role in the optimisation. In the frequency domain design stage, a multisine signal is produced. In the time domain design stage, the

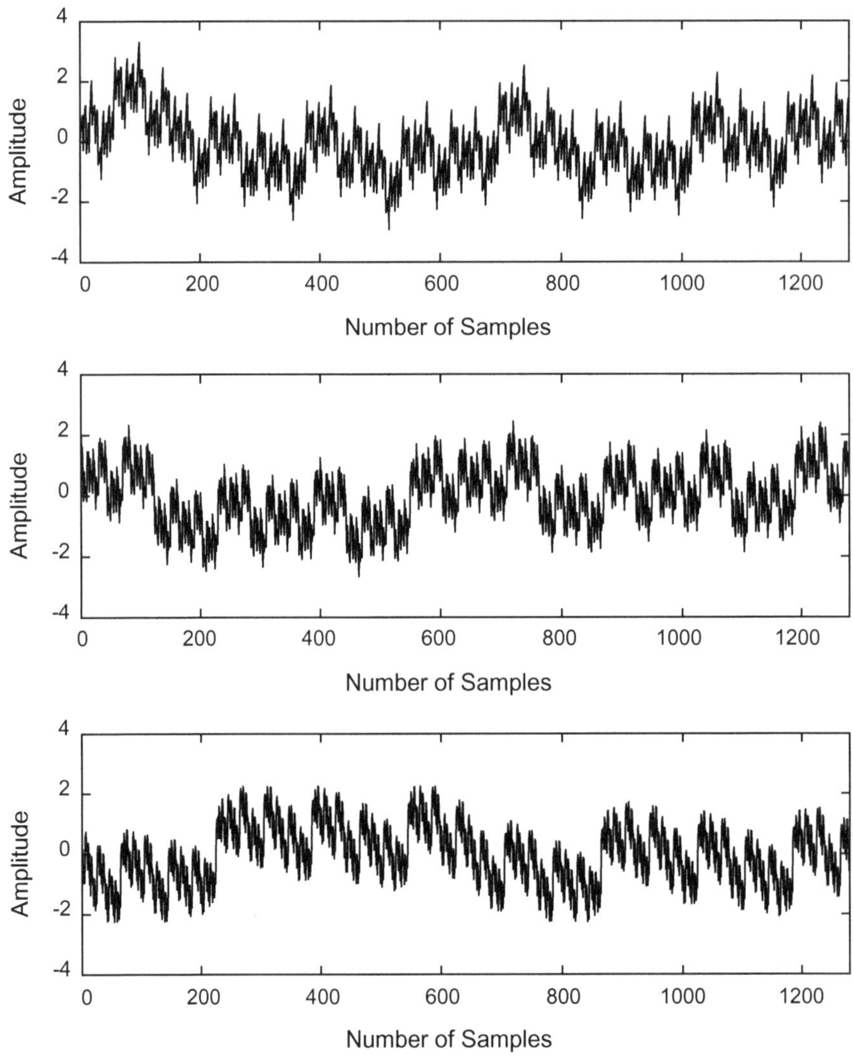

Fig. 3.7 Multisine signals for Specification D without ZOH pre-compensation. Top: Schroeder; middle: time-frequency swapping; bottom: infinity norm

multisine signal is quantised into a specified number of levels M. The mapping in the time domain is given by

$$Q_i = \left| \frac{B_i}{q^{i-1}} - \frac{B_{i+1}}{q^i} \right| \begin{array}{c} i = 1, \ldots, M/2\ M \text{ even} \\ i = 1, \ldots, (M-1)/2\ M \text{ odd} \end{array}, \tag{3.8}$$

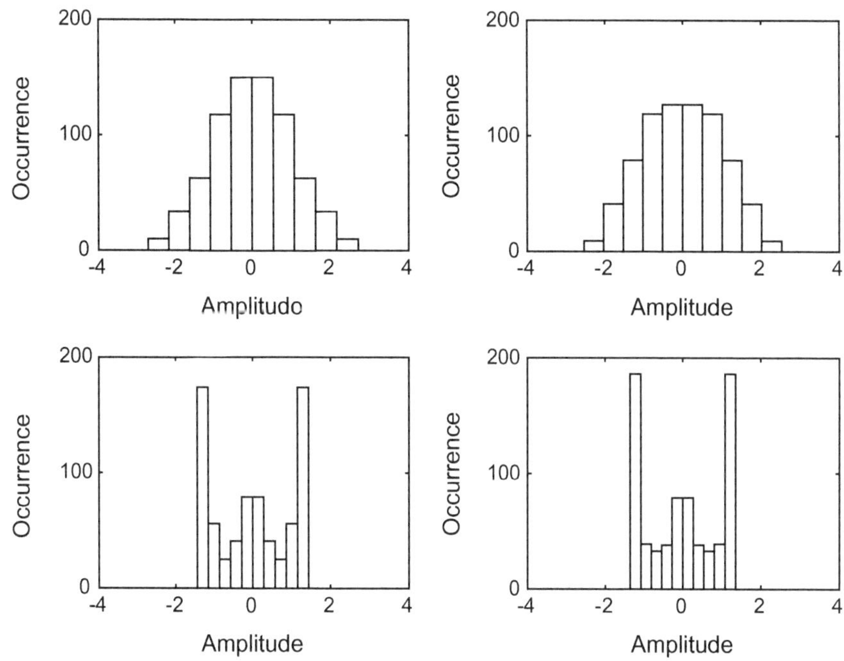

Fig. 3.8 Histograms showing the amplitude distribution of the multisine for Specification C with ZOH pre-compensation. Top, left: random-phase multisine; top, right: infinity norm with 2 iterations up to $p = 4$; bottom, left: infinity norm with 50 iterations up to $p = 4$; bottom, right: infinity norm with 50 iterations up to $p = 256$

Table 3.4 Performance measures of DIB and DIT signals generated without ZOH pre-compensation

Specification	Signal type	Crest factor	PIPS (%)	P_{uf} (%)
A	DIB	1	99.97	92.99
A	DIT	1.07	93.79	86.58
B	DIB	1	100	94.68
B	DIT	1.20	83.19	67.42
C	DIB	N/A	N/A	N/A
C	DIT	1.29	77.29	57.75
D	DIB	1	99.95	68.47
D	DIT	1	99.97	66.29

where B_i, $i = 1, 2, \ldots, M + 1$, results from dividing the input range with maximum amplitude u_{max} and minimum amplitude u_{min} into M equal segments (McCormack et al. 1995; Tan and Godfrey 2004). In particular,

$$B_i = u_{min} + \left(\frac{u_{max} - u_{min}}{M}\right) \times (i - 1). \tag{3.9}$$

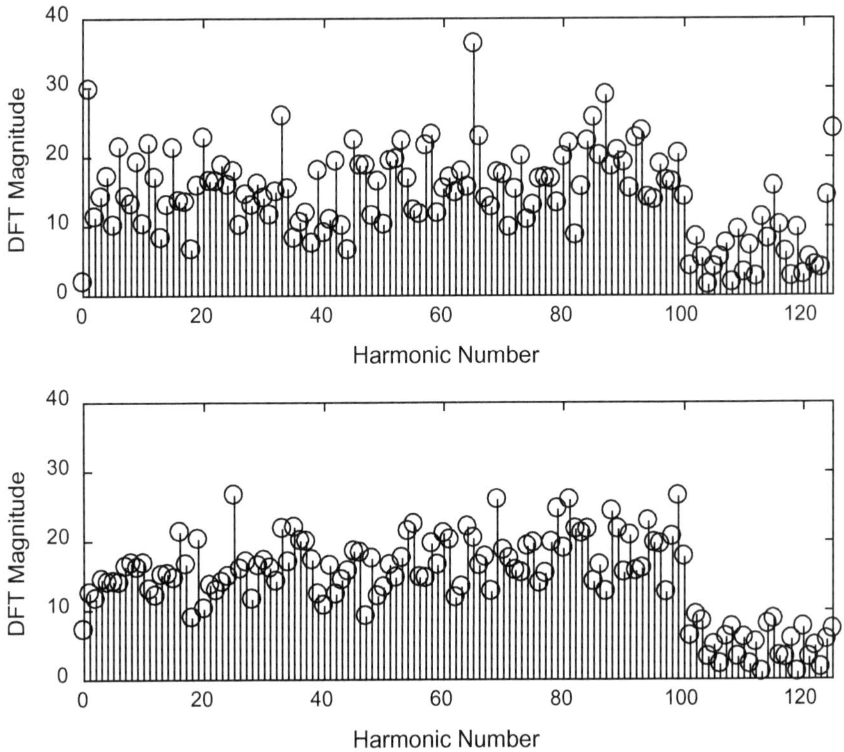

Fig. 3.9 DFT magnitudes obtained for Specification A with ZOH pre-compensation. Top: DIB signal; bottom: DIT signal

Note that the input bands are symmetrical about zero. The quantiser parameter q is very important when $M > 2$ as it determines the amplitude distribution of the resulting signal.

Example

Assume that an input signal with amplitude ranging from $u_{\min} = -1$ to $u_{\max} = 1$ is to be quantised into a three-level signal. Plot the characteristic line of the quantiser for $q = 1$, 2 and 4. Repeat for a four-level signal.

Solution

First, consider the case for quantisation into a three-level signal ($M = 3$). We have $B_1 = -1$, $B_2 = -0.333$, $B_3 = 0.333$ and $B_4 = 1$. When $q = 1$, $Q_1 = \left|\frac{B_1}{q^0} - \frac{B_2}{q^1}\right| = \left|-1 - \left(\frac{-0.333}{1}\right)\right| = 0.667$. This means that the first and also the third (which is the last) input bands are of size 0.667 due to symmetricity of the quantiser. As a result, the second (middle) band is also of size 0.667 and all the bands are of equal size. When $q = 2$, $Q_1 = \left|-1 - \left(\frac{-0.333}{2}\right)\right| = 0.833$. The first and third bands are of size 0.833

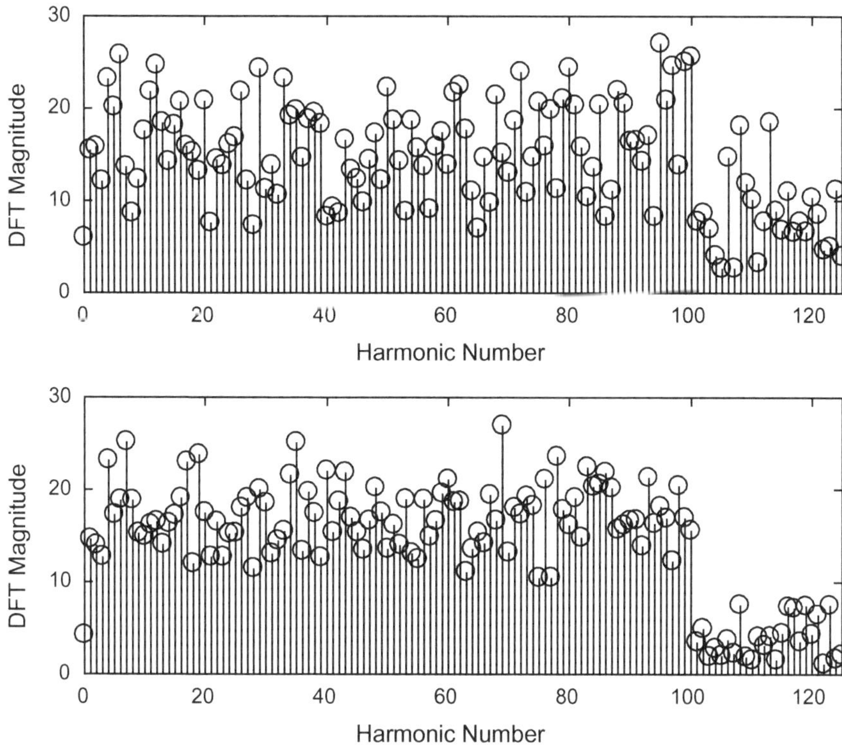

Fig. 3.10 DFT magnitudes obtained for Specification A without ZOH pre-compensation. Top: DIB signal; bottom: DIT signal

and the second band is of size 0.333. When $q = 4$, $Q_1 = \left|-1 - \left(\frac{-0.333}{4}\right)\right| = 0.917$. The first and third bands are of size 0.917 and the second band is of size 0.167. The characteristic line of the quantiser is illustrated in Fig. 3.11, where it can be seen that as q increases, a larger input band is mapped into the levels ± 1. As a consequence of this, when a three-level MLMH signal is generated with a large value of q, the resulting signal becomes near binary.

Next, consider the case for quantisation into a four-level signal ($M = 4$). Now $B_1 = -1$, $B_2 = -0.5$, $B_3 = 0$, $B_4 = 0.5$ and $B_5 = 1$. When $q = 1$, $Q_1 = \left|\frac{B_1}{q^0} - \frac{B_2}{q^1}\right| = \left|-1 - \left(\frac{-0.5}{1}\right)\right| = 0.5$ and $Q_2 = \left|\frac{B_2}{q^1} - \frac{B_3}{q^2}\right| = \left|-\frac{0.5}{1} - 0\right| = 0.5$. The input bands are all of equal size. When $q = 2$, $Q_1 = \left|-1 - \left(\frac{-0.5}{2}\right)\right| = 0.75$ and $Q_2 = \left|-\frac{0.5}{2} - 0\right| = 0.25$. The first and the fourth (last) input bands are larger than the second and third ones. This difference is even more pronounced when $q = 4$. In this case, $Q_1 = \left|-1 - \left(\frac{-0.5}{4}\right)\right| = 0.875$ and $Q_2 = \left|-\frac{0.5}{4} - 0\right| = 0.125$. The characteristic line of the quantiser is shown in Fig. 3.12.

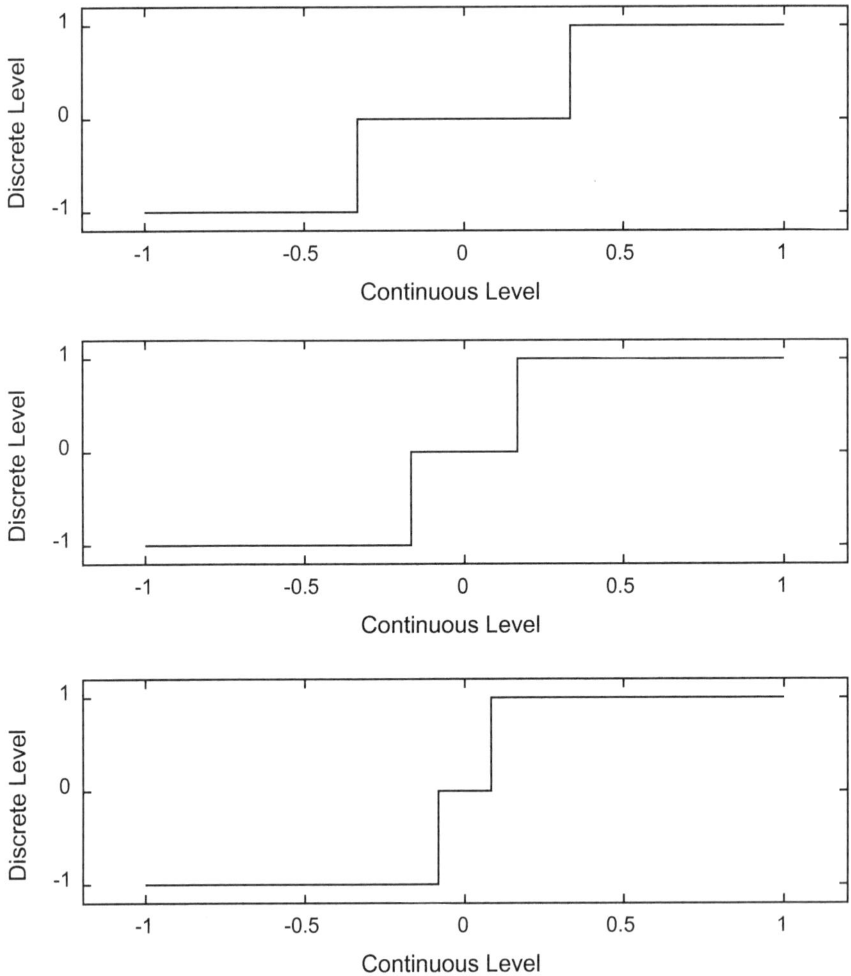

Fig. 3.11 Characteristic curve of quantiser for $M = 3$ with $q = 1$ (top), $q = 2$ (middle) and $q = 4$ (bottom)

In the `multilev_new` algorithm, the value of q is chosen based on the best signal produced when Schroeder phases are applied. For this value of q, random phases are also tested as initial values during different trials. The number of trials can be specified by the user. The performance measures for Specifications A to D (see Sect. 3.1) are summarised in Tables 3.5 and 3.6. The default number of trials of 10,000 was applied. Despite the large number of trials, the optimisation took only a few seconds. The signals were then scaled to have an RMS value of 1. The plots for Specification A are shown in Figs. 3.13, 3.14, 3.15 and 3.16.

From Tables 3.5 and 3.6, a general trend is observed that as the number of signal levels increases, the values of PIPS and PIPSE decrease leading to higher crest factor.

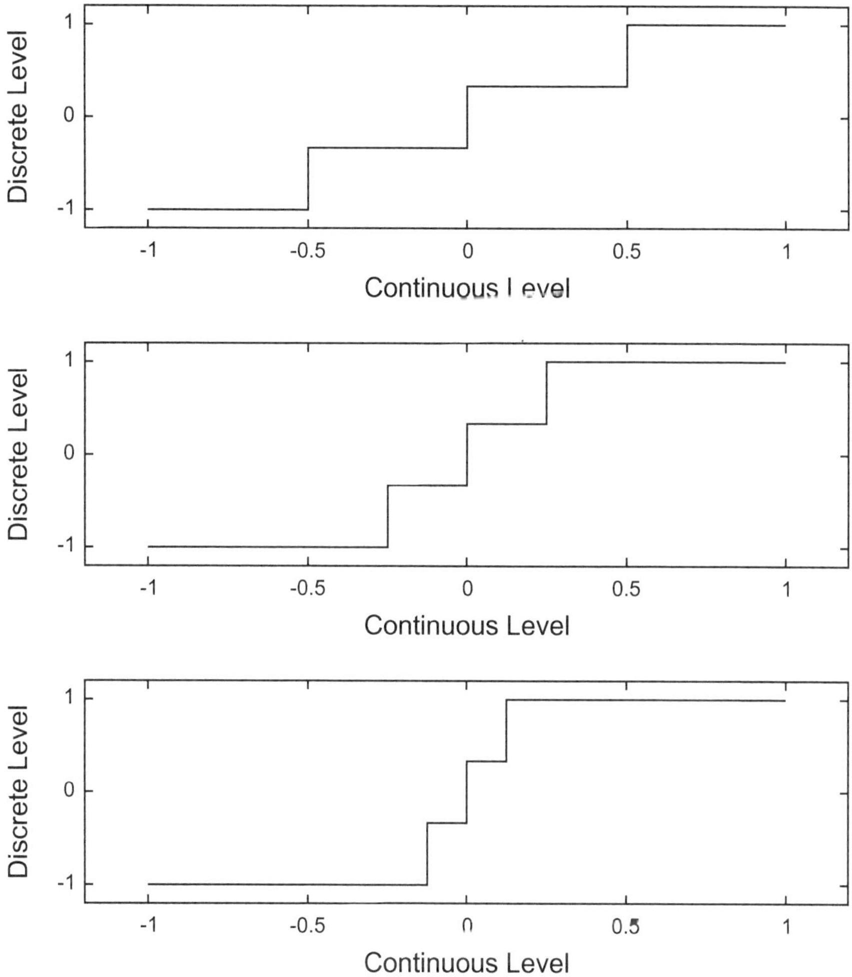

Fig. 3.12 Characteristic curve of quantiser for $M = 4$ with $q = 1$ (top), $q = 2$ (middle) and $q = 4$ (bottom)

However, the uniformity of the excited harmonics typically improves with a larger number of signal levels. This can be observed from the value of EMINE which increases with the number of signal levels for all the Specifications A to D, as well as from Figs. 3.14 and 3.16.

It is interesting also to compare the performance measures of MLMH signals with $M = 2$ and 3, with those of the corresponding DIB and DIT signals. In particular, EMINE values are typically higher for MLMH signals (Table 3.5) compared with DIB and DIT signals (Table 3.3). This is likely due to EMINE being optimised

Table 3.5 Performance measures of MLMH signals generated with ZOH pre-compensation

Specification	Number of signal levels	Crest factor	PIPS (%)	PIPSE (%)	EMINE (%)
A	2	1	100	90.06	51.99
A	3	1.07	93.59	84.08	69.56
A	5	1.10	91.10	82.30	77.85
B	2	1	100	88.75	51.87
B	3	1.18	84.62	76.68	71.96
B	5	1.19	83.96	75.71	81.49
C	2	N/A	N/A	N/A	N/A
C	3	1.22	81.65	72.90	56.95
C	5	1.29	77.46	69.92	74.02
D	2	1	99.99	78.76	98.41
D	3	1.08	92.15	76.98	98.56
D	5	1.13	88.44	75.66	98.61

Table 3.6 Performance measures of MLMH signals generated without ZOH pre-compensation

Specification	Number of signal levels	Crest factor	PIPS (%)
A	2	1	100
A	3	1.03	97.33
A	5	1.17	85.60
B	2	1	100
B	3	1.11	89.89
B	5	1.18	84.85
C	2	N/A	N/A
C	3	1.33	75.54
C	5	1.31	76.42
D	2	1	99.99
D	3	1.12	88.87
D	5	1.12	88.87

directly in `multilev_new`. However, no clear trend is observed for the values of PIPS and PIPSE.

The changes of the performance measures with the quantiser parameter q are shown in Figs. 3.17 and 3.18 for Specification A. As q increases, a larger input band is quantised to larger output values resulting in the signal having more values closer to the maximum and minimum amplitudes. Thus PIPS and PIPSE both show increasing trends while the crest factor shows an opposite trend. EMINE also decreases with q despite an increasing trend for small values of q. The plots for Specifications B and C exhibit the same pattern and are not shown here.

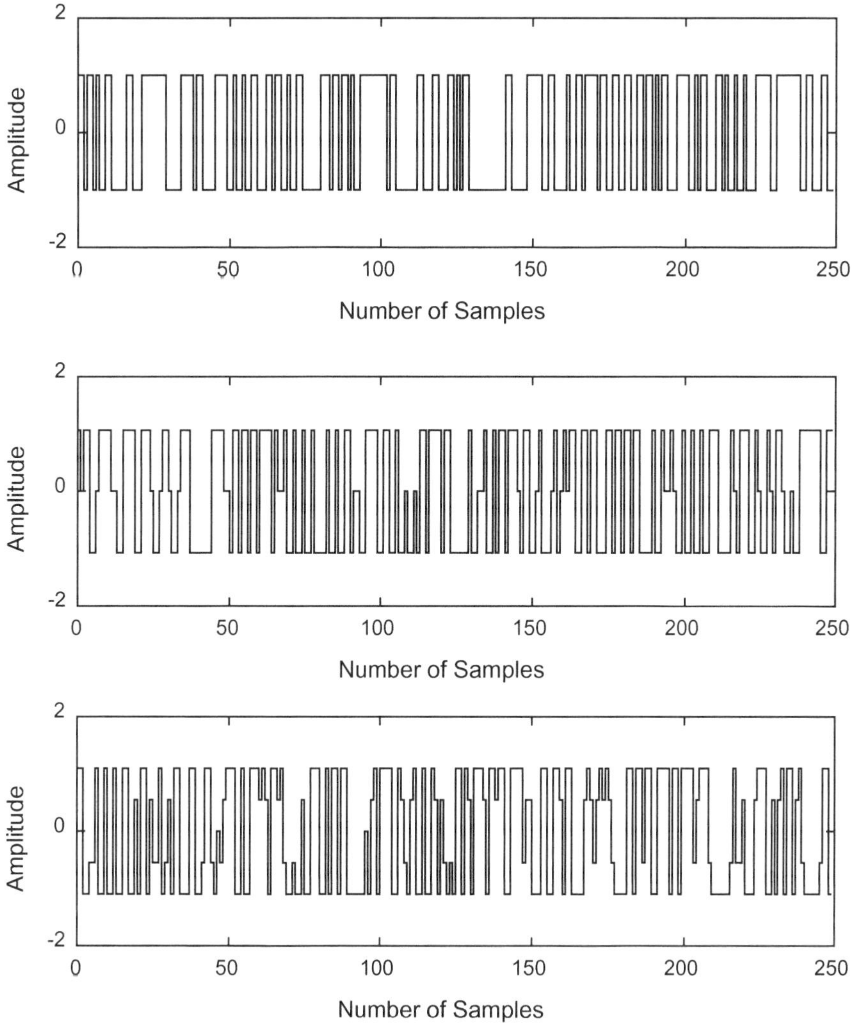

Fig. 3.13 MLMH signals for Specification A with ZOH pre-compensation. Top: $M = 2$; middle: $M = 3$; bottom: $M = 5$

However, for Specification D as depicted in Figs. 3.19 and 3.20, the changes in the performance measures are smoother than those obtained for Specifications A to C. In addition to this, EMINE stays close to 100% throughout. This is due to Specification D having only 10 excited harmonics and hence, it is easier for the optimisation programme to ensure sufficient power in these harmonics.

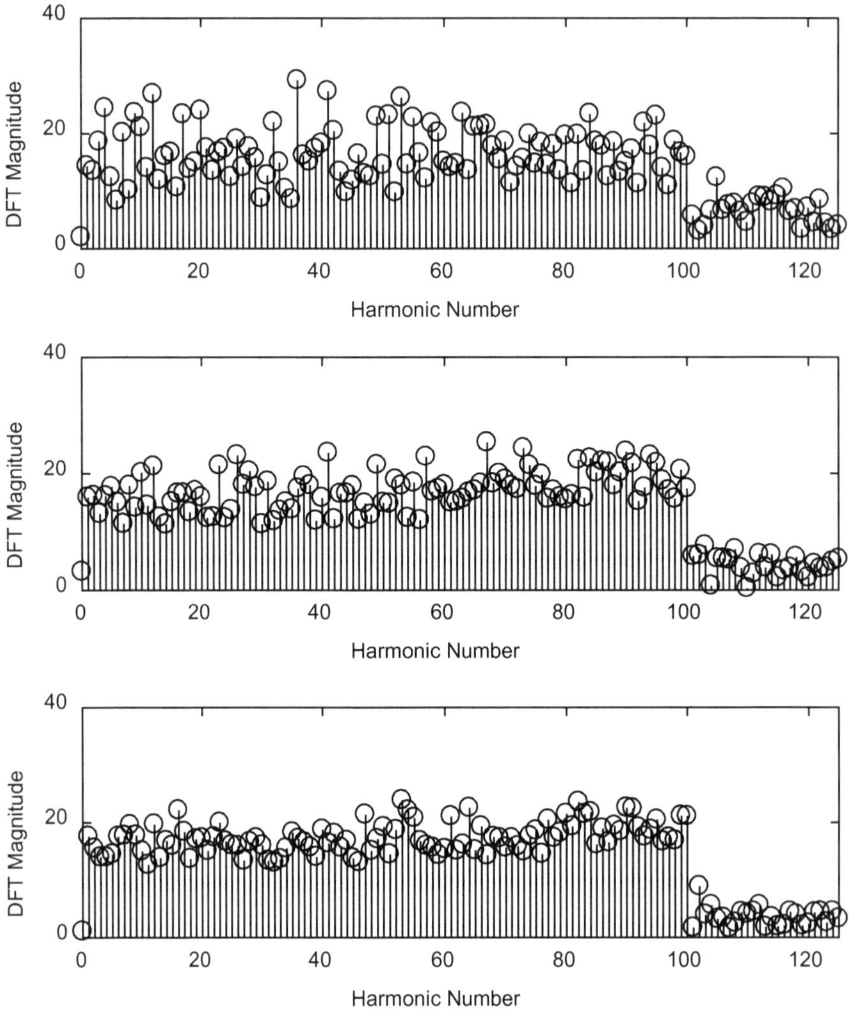

Fig. 3.14 DFT magnitudes obtained using MLMH signals for Specification A with ZOH pre-compensation. Top: $M = 2$; middle: $M = 3$; bottom: $M = 5$

3.4 Hybrid Signals

Hybrid signals are formed using a combination of pseudorandom and computer-optimised techniques. An example is the Galois-multilev signals, or in short, Gallev signals (Tan et al. 2005) which are generated following the design of PRML signals with an important design step carried out using the `multilev_new` algorithm.

The steps for designing a Gallev signal are as follows:

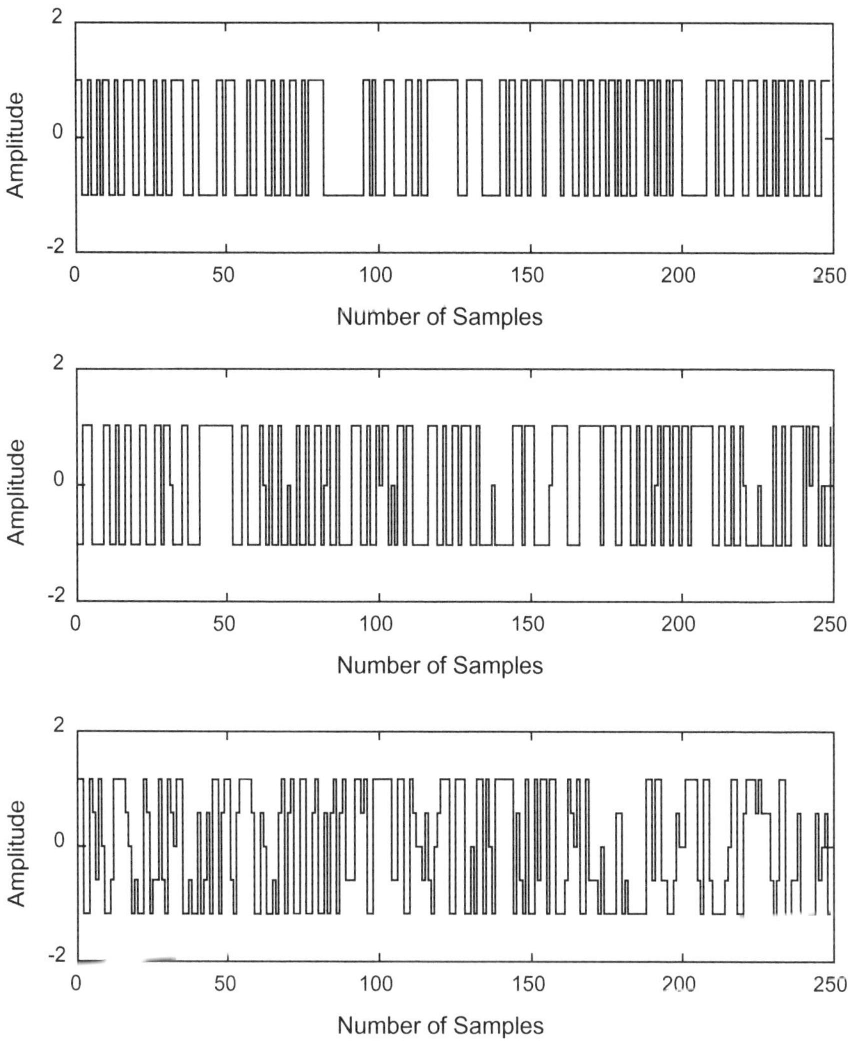

Fig. 3.15 MLMH signals for Specification A without ZOH pre-compensation. Top: $M = 2$; middle: $M = 3$; bottom: $M = 5$

Step 1: Depending on the application, specify the required harmonic properties and select a range of suitable periods N for the Gallev signal. Decide also on the number of signal levels required.
Step 2: Choose suitable values of q and n given that the period $N = q^n - 1$ for a PRML signal generated from GF(q). Refer to Sect. 2.3.1 for more details on the design of PRML signals.

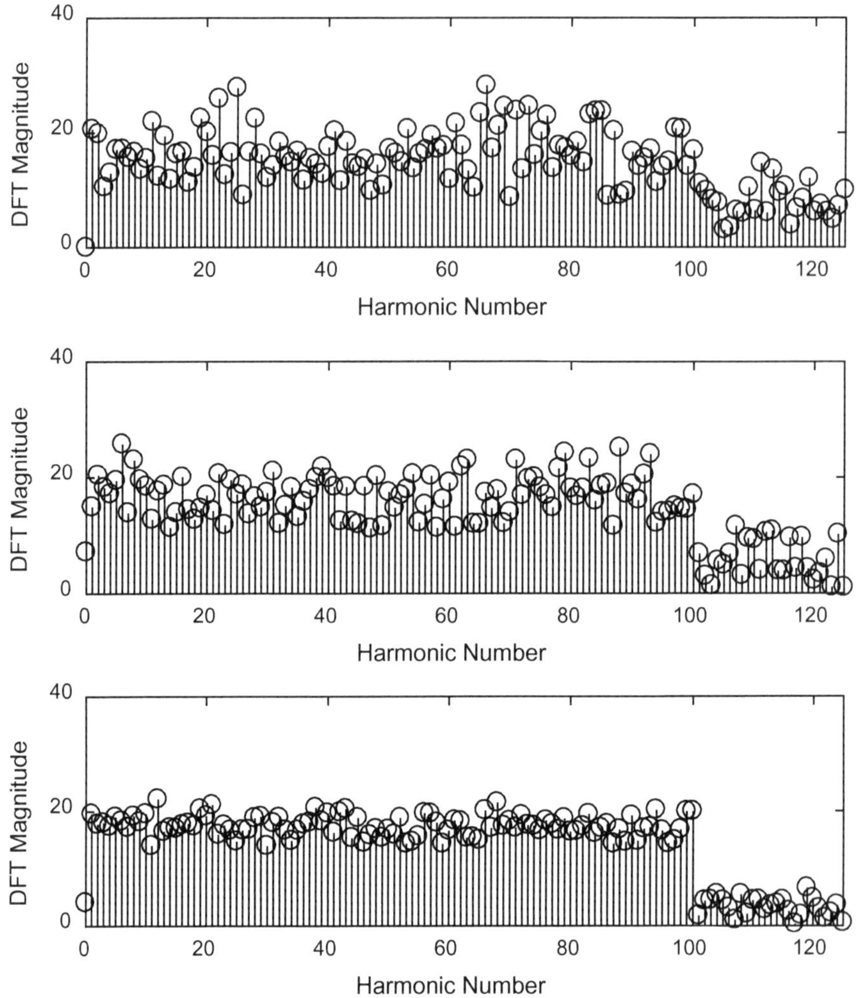

Fig. 3.16 DFT magnitudes obtained using MLMH signals for Specification A without ZOH pre-compensation. Top: $M = 2$; middle: $M = 3$; bottom: $M = 5$

Step 3: Generate a primitive signal $u_{q,1}(i)$ of length $q - 1$ from GF(q) with the desired harmonic properties as an MLMH signal (see Sect. 3.3). Use the `multilev_new` algorithm for the optimisation.
Step 4: Obtain the sequence-to-signal conversion by comparing the primitive signal $u_{q,1}(i)$ with the primitive sequence $s_{q,1}(i) = [\,1\ g\ g^2\ g^3\ \ldots\ g^{q-2}]$ modulo-q. Further to this, the sequence value 0 is converted into signal level 0.
Step 5: Generate a PRML sequence $s_{q,n}(i)$ based on Eq. 2.20.

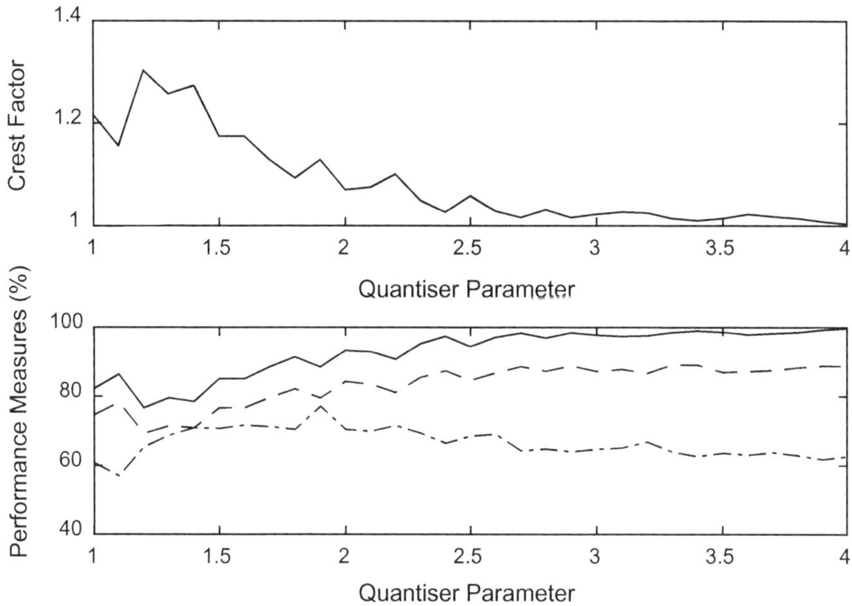

Fig. 3.17 Variation of crest factor and other performance measures for Specification A with ZOH pre-compensation and $M = 3$. In the bottom plot, the measures are PIPS (solid line), PIPSE (dashed line) and EMINE (dashed-dotted line)

Step 6: Convert the sequence $s_{q,n}(i)$ into a signal $u_{q,n}(i)$ using the sequence-to-signal conversion obtained in Step 4. The resulting signal $u_{q,n}(i)$ is a Gallev signal.

Gallev signals are particularly suitable when the required signal has a large number of harmonics specified and where the excited harmonics are not required to be strictly uniform. In such cases, due to the long period $N = q^n - 1$, the generation of MLMH signals may be comparatively slow. However, the use of Gallev allows the MLMH optimisation to be carried out on primitive signals with a much shorter period $q - 1$. This capitalises on the fact that the primitive signal $u_{q,1}(i)$ has DFT magnitude which is closely related that of $u_{q,n}(i)$ with $n \neq 1$, according to Eq. 2.22.

Example

Generate a five-level Gallev signal with $N = 6858$ having harmonic multiples of two and three suppressed.

Solution

This can be achieved using $q = 19$ and $n = 3$ since $19^3 - 1 = 6858$. Note that q must be at least equal to 5 to generate a five-level signal. However, the requirement of having harmonic multiples of two and three suppressed imposes the constraint that $q - 1$ must be an integer multiple of 6. The primitive signal $u_{19,1}(i)$ is given by

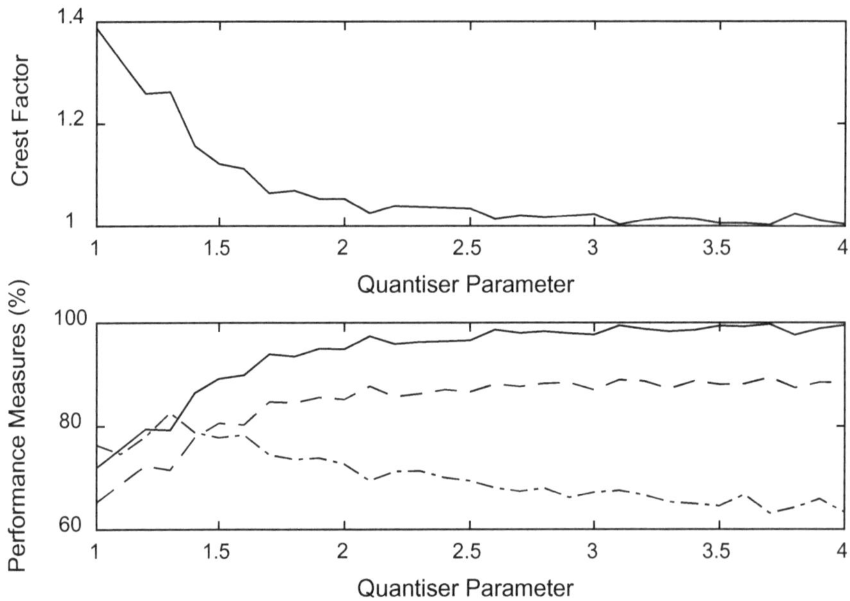

Fig. 3.18 Variation of crest factor and other performance measures for Specification A with ZOH pre-compensation and $M = 5$. In the bottom plot, the measures are PIPS (solid line), PIPSE (dashed line) and EMINE (dashed-dotted line)

$$u_{19,1}(i) = [2\ -1\ -1\ 2\ -2\ -2\ 0\ -1\ -1\ -2\ 1\ 1\ -2\ 2\ 2\ 0\ 1\ 1\ 1] \tag{3.10}$$

and its DFT magnitude is shown in Fig. 3.21. The primitive sequence using primitive element $g = 2$ is

$$s_{19,1}(i) = [1\ 2\ 4\ 8\ 16\ 13\ 7\ 14\ 9\ 18\ 17\ 15\ 11\ 3\ 6\ 12\ 5\ 10]. \tag{3.11}$$

Comparing Eqs. 3.10 and 3.11, the sequence-to-signal conversion is given in Table 3.7. A PRML sequence $s_{19,3}(i)$ is generated and converted into a signal $u_{19,3}(i)$ using the sequence-to-signal conversion obtained. The resulting signal is a Gallev signal with a PIPS value of 72.55%.

In contrast, a five-level PRML signal with harmonic multiples of two and three suppressed generated directly from GF(19) has a PIPS value of 39.74%. The trade-off is that the latter has uniformly excited harmonics whereas the Gallev signal, in the quest for higher PIPS, compromises the uniformity of the excited harmonics. The DFT magnitude of the Gallev signal is shown in Fig. 3.22 where the non-uniformity in the harmonics is seen to follow the same pattern as that of the primitive signal shown in Fig. 3.21.

The primitive signal $u_{q,1}(i)$ can also be designed to have subperiods within a period $q - 1$, in which case the final Gallev signal will have a period shorter than $q^n - 1$,

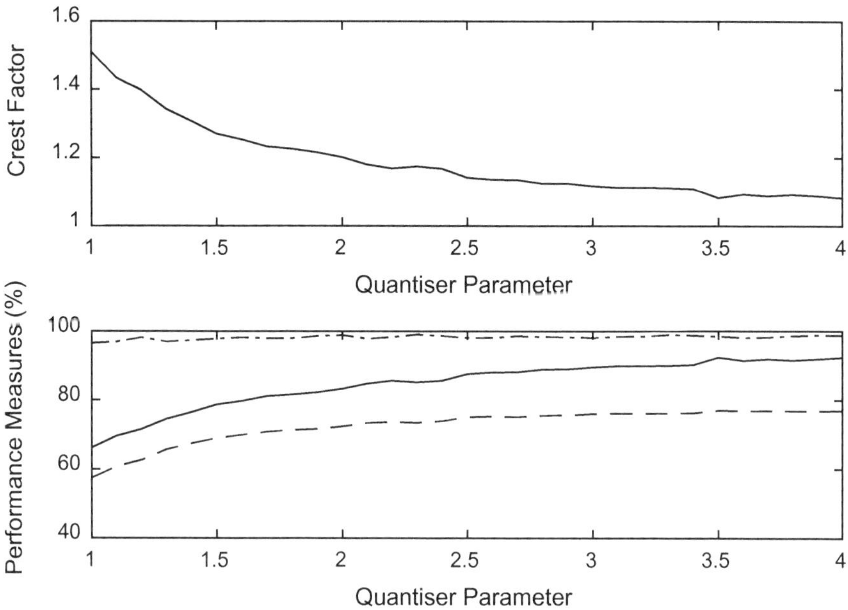

Fig. 3.19 Variation of crest factor and other performance measures for Specification D with ZOH pre-compensation and $M = 3$. In the bottom plot, the measures are PIPS (solid line), PIPSE (dashed line) and EMINE (dashed-dotted line)

but which is a factor of $q^n - 1$. This follows along the same principles as for the design of truncated PRML signals (see Sect. 2.3.2).

Example

Generate a five-level Gallev signal with $N = 684$ having harmonic multiples of two and three suppressed.

Solution

This can be achieved using $q = 37$ and $n = 2$. The primitive signal $u_{37,1}(i)$ (of length 36) is formed by concatenating $u_{19,1}(i)$ (of length 18) in Eq. 3.10 twice. Thus, $u_{37,1}(i)$ is given by

$$\begin{aligned} u_{37,1}(i) = [&2\ -1\ -1\ 2\ -2\ -2\ 0\ -1\ -1\ -2\ 1\ 1\ -2\ 2\ 2\ 0\ 1\ 1 \\ &2\ -1\ -1\ 2\ -2\ -2\ 0\ -1\ -1\ -2\ 1\ 1\ -2\ 2\ 2\ 0\ 1\ 1] \end{aligned} \tag{3.12}$$

and its DFT magnitude is shown in Fig. 3.23. The primitive sequence using primitive element $g = 2$ is

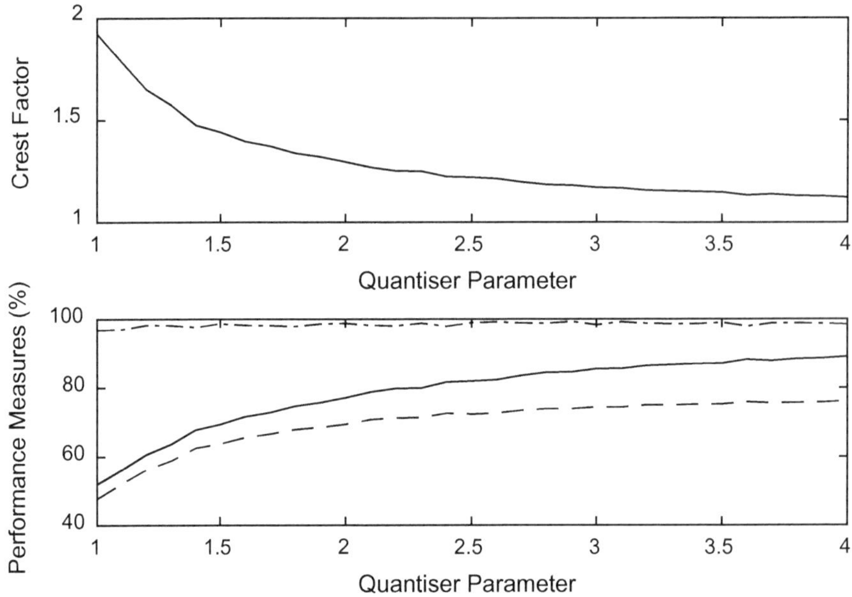

Fig. 3.20 Variation of crest factor and other performance measures for Specification D with ZOH pre-compensation and $M = 5$. In the bottom plot, the measures are PIPS (solid line), PIPSE (dashed line) and EMINE (dashed-dotted line)

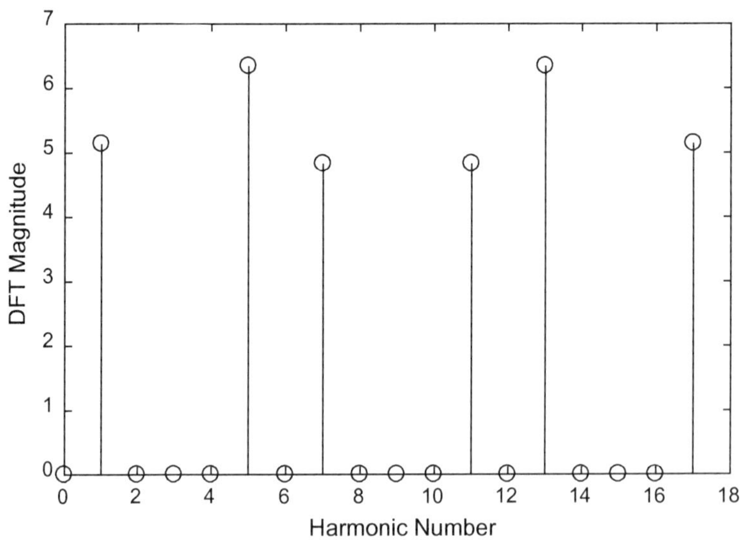

Fig. 3.21 DFT magnitude of the primitive signal generated from GF(19)

Table 3.7 Sequence-to-signal conversion for a Gallev signal from GF(19)

Sequence value (field element)	Signal level
0	0
1	2
2	−1
3	2
4	−1
5	1
6	2
7	0
8	2
9	−1
10	1
11	−2
12	0
13	−2
14	−1
15	1
16	−2
17	1
18	−2

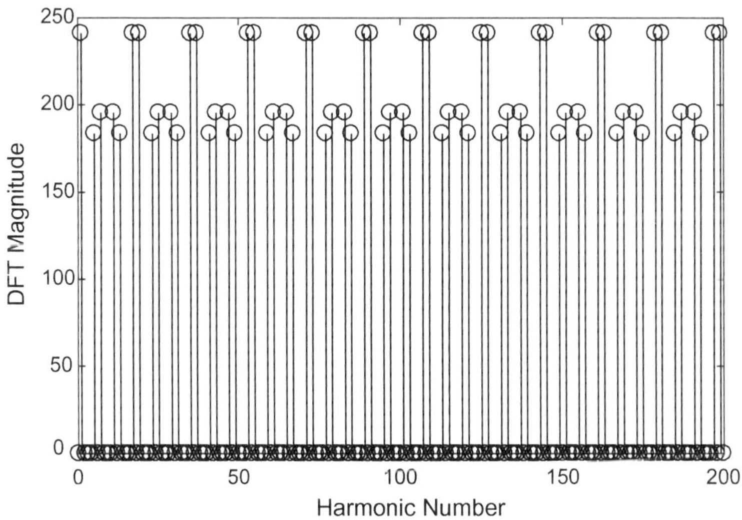

Fig. 3.22 DFT magnitude of the Gallev signal generated from GF(19) with characteristic equation $4D^3 \oplus_{19} D^2 \oplus_{19} 1 = 0$ and $N = 6858$, showing only the harmonics up to 200 for the sake of clarity

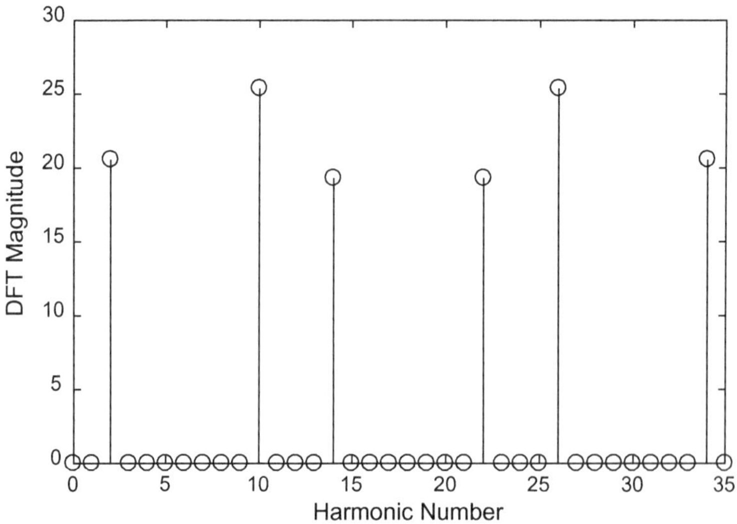

Fig. 3.23 DFT magnitude of the primitive signal generated from GF(37)

$$s_{37,1}(i) = [\,1\ 2\ 4\ 8\ 16\ 32\ 27\ 17\ 34\ 31\ 25\ 13\ 26\ 15\ 30\ 23\ 9\ 18\ 36\ 35\ 33\ 29\ 21\ 5\ 10\ 20\ 3\ 6\ 12\ 24\ 11\ 22\ 7\ 14\ 28\ 19\,]. \quad (3.13)$$

Comparing Eqs. 3.12 and 3.13, the sequence-to-signal conversion is given in Table 3.8. A PRML sequence $s_{37,2}(i)$ is generated and converted into a signal $u_{37,2}(i)$ using the sequence-to-signal conversion obtained. However, since $u_{37,1}(i)$ contains two subperiods within a period of length $q - 1 = 37 - 1 = 36$, the signal $u_{37,2}(i)$ also contains two subperiods within a period of length $q^n - 1 = 37^2 - 1 = 1368$. Hence, $u_{37,2}(i)$ is truncated to $N = 1368/2 = 684$. The resulting signal is a Gallev signal with a PIPS value of 73.55%. The DFT magnitude of the Gallev signal is shown in Fig. 3.24.

3.5 Optimal Input Signals

Optimal signals are designed to minimise an application cost function. In a control setting, the application cost function may be related to the degradation in the control performance due to uncertainty in the model estimate. The objective is to find the input spectrum Φ_u that minimises the experimental cost f_{cost} subject to the estimated parameters giving a model with acceptable application performance (Annergren et al. 2017). This can be stated as

Table 3.8 Sequence-to-signal conversion for a Gallev signal from GF(37)

Sequence value (field element)	Signal level
0	0
1	2
2	−1
3	−1
4	−1
5	−2
6	−2
7	2
8	2
9	1
10	0
11	−2
12	1
13	1
14	0
15	2
16	−2
17	−1
18	1
19	1
20	−1
21	−2
22	2
23	0
24	1
25	1
26	2
27	0
28	1
29	2
30	2
31	−2
32	−2
33	−1
34	−1
35	−1
36	2

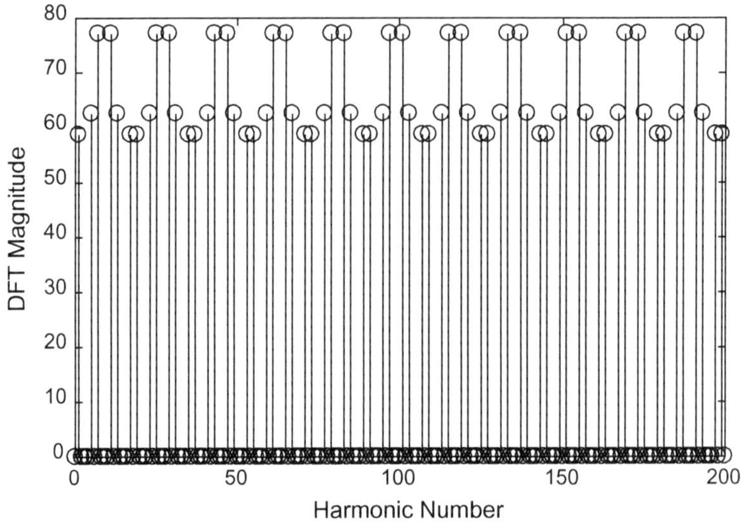

Fig. 3.24 DFT magnitude of the Gallev signal generated from GF(37) with characteristic equation $5D^2 \oplus_{37} D \oplus_{37} 1 = 0$ and $N = 684$, showing only the harmonics up to 200 for the sake of clarity

$$\underset{\Phi_u}{\text{minimise}}\; f_{\text{cost}}(\Phi_u) \tag{3.14}$$

subject to

$$\varepsilon_{\text{SI}}(\alpha) \subseteq \theta_{\text{app}}(\gamma) \tag{3.15}$$

$$\Phi_u(\omega) \text{ is positive semidefinite } \forall\omega. \tag{3.16}$$

The condition in Eq. 3.15 is interpreted as follows. The prediction error method guarantees, under weak assumptions and asymptotically in the number of observations N (which can also be seen as the signal period), with probability α that the obtained estimates lie inside a particular ellipsoid ε_{SI} centred around the true parameters. In contrast, θ_{app} is the set of parameters that result in acceptable application performance. This set depends on the allowable degradation in performance defined by γ.

The design requires some a priori knowledge of the system. This can be obtained for example, from broadband identification tests or from step tests. The design may require iteration depending on the application. When iteration is applied, the model is improved in every iteration leading to an improved perturbation signal. This signal further results in better estimates of the model.

Once the spectrum has been optimised according to Eq. 3.14, it needs to be converted to a time realisation of the signal. According to Wahlberg et al. (2010), there are two direct ways to find a time realisation for a given optimised spectrum—the first one is to use an autoregressive process and the second one is to use sinusoidal signals in white noise.

Another way of optimising the input power spectrum is by making use of the dispersion function $\nu(\chi, \Omega_k)$ with $\Omega_k = \mathrm{j}\omega_k$ for continuous-time systems and $\Omega_k = \mathrm{e}^{-\mathrm{j}\omega_k t_s}$ for discrete-time systems where ω_k denotes the angular frequency at harmonic k and t_s is the sampling interval (Pintelon and Schoukens 2012). The dispersion function $\nu(\chi, \Omega_k)$ for a given power spectrum $\chi(\Omega) = (|U(1)|^2 \ldots |U(F)|^2)$ with $\sum_{k=1}^{F} |U(k)|^2 = \wp$ is given by Pintelon and Schoukens (2012)

$$\nu(\chi, \Omega_k) = \mathrm{trace}([Fi(\chi)]^{-1} fi(\Omega_k)) \tag{3.17}$$

with $Fi(\chi)$ denoting the information matrix resulting from the design $\chi(\Omega)$ and $fi(\Omega_k)$ denoting the information matrix corresponding to a single frequency input with a normalised power spectrum $|U(k)|^2 = \wp$.

The optimal input can be obtained through an iterative algorithm which starts with an input spectrum where equal power is allocated to all the F excited harmonics (Pintelon and Schoukens 2012). The design is iterated using

$$\chi_{i+1}(\Omega_k) = \chi_i(\Omega_k)\nu(\chi_i, \Omega_k)/n_\theta, k = 1, 2, \ldots F, \tag{3.18}$$

where i denotes the iteration number and n_θ is the number of unknown model parameters.

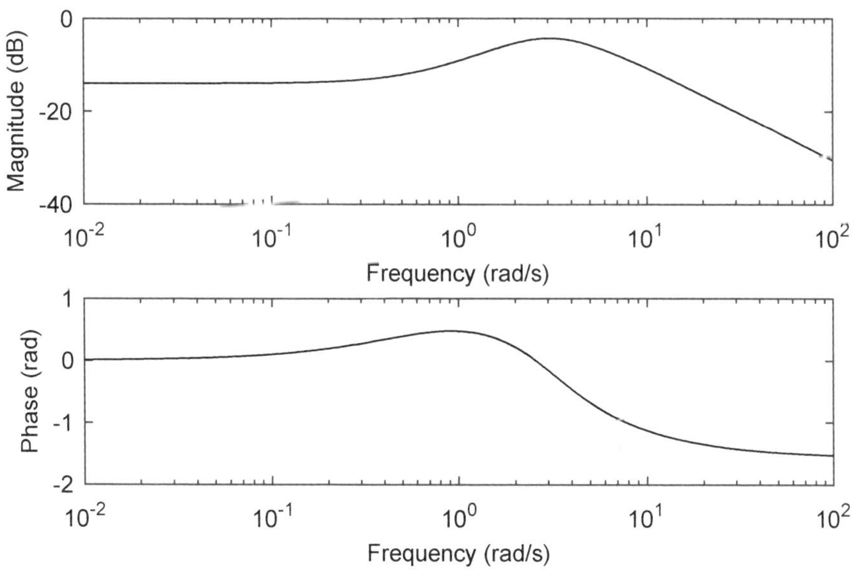

Fig. 3.25 Bode plot of the system given by Eq. 3.19

As an example, the design of optimal power spectrum is illustrated for a system with transfer function

$$G(s) = \frac{3s + 2}{s^2 + 5s + 10}. \tag{3.19}$$

In the simulation, it was assumed that all the system parameters (numerator and denominator coefficients in Eq. 3.19) were unknown, with $n_\theta = 5$. The excited frequencies were set to 0.1, 0.2, 0.3, ..., 2 Hz, giving $F = 20$. The Bode plot of the

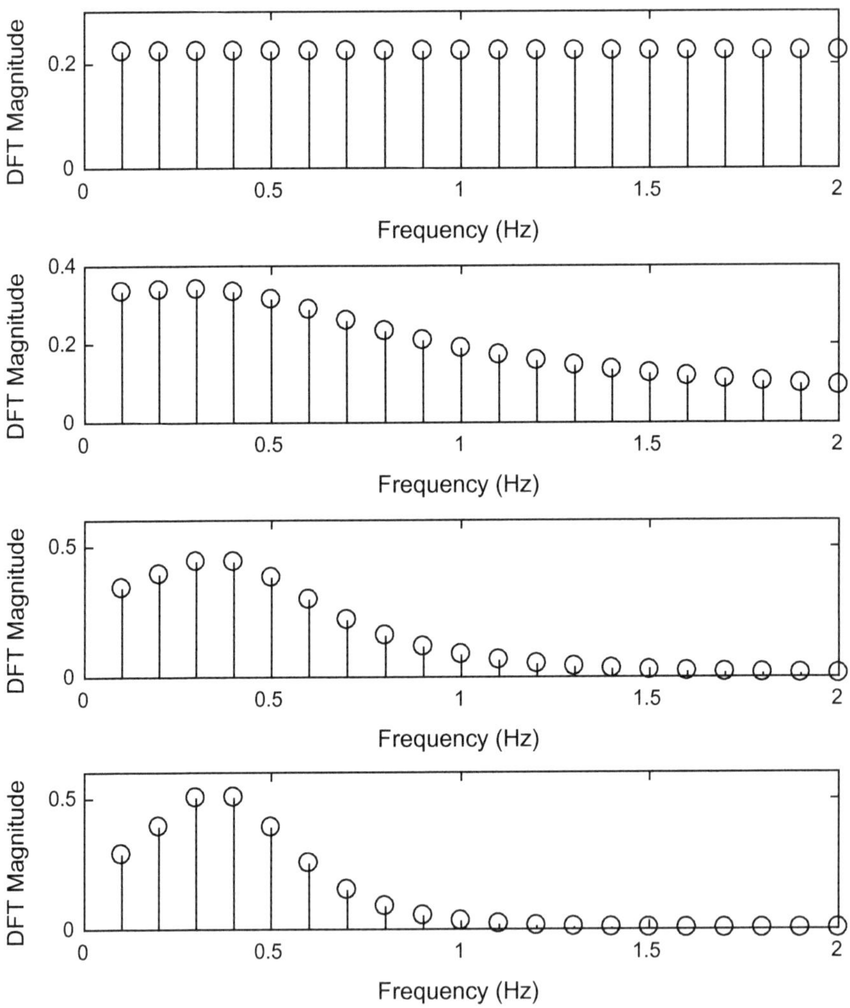

Fig. 3.26 Evolution of the optimal input design. From top to bottom: initial design, design after 1 iteration, design after 3 iterations, design after 5 iterations

system is shown in Fig. 3.25, with the resonant peak appearing at around 3 rad s^{-1} (0.4775 Hz). Comparing with the Bode plot, the set of excited frequencies is in fact not very well selected, but the aim here is to illustrate what the algorithm will do to improve the design. The evolution of the optimal input is shown in Fig. 3.26. The algorithm gradually assigns more power into the frequencies close to the resonant peak.

References

Annergren M, Larsson CA, Hjalmarsson H, Bombois X, Wahlberg B (2017) Application-oriented input design in system identification: optimal input design for control. IEEE Control Syst Mag 37:31–56

Boyd S (1986) Multitone signals with low crest factor. IEEE Trans Circ Syst 33:1018–1022

Cham CL, Tan AH, Tan WH (2017) Identification of a multivariable nonlinear and time-varying mist reactor system. Control Eng Pract 63:13–23

Faifer M, Ottoboni R, Toscani S, Cherbaucich C, Mazza P (2015) Metrological characterization of a signal generator for the testing of medium-voltage measurement transducers. IEEE Trans Instrum Meas 64:1837–1846

Gersho A, Gopinath B, Odlyzko AM (1979) Coefficient inaccuracy in transversal filtering. Bell Syst Technol J 58:2301–2316

Godfrey KR, Tan AH, Barker HA, Chong B (2005) A survey of readily accessible perturbation signals for system identification in the frequency domain. Control Eng Pract 13:1391–1402

Guillaume P, Schoukens J, Pintelon R, Kollár I (1991) Crest-factor minimization using nonlinear Chebyshev approximation methods. IEEE Trans Instrum Meas 40:982–989

Kazazis S, Esterer N, Depalle P, McAdams S (2017) A performance evaluation of the Timbre Toolbox and the MIRtoolbox on calibrated test sounds. In: Proceedings of the international symposium on musical acoustics, Montreal, Canada, 18–22 June, pp 144–147

Kollár I (1994) Frequency domain system identification toolbox for use with MATLAB. The Math-Works Inc., Natick, MA

Kulesza Z (2014) Dynamic behaviour of cracked rotor subjected to multisine excitation. J Sound Vib 333:1369–1378

McCormack AS, Godfrey KR, Flower JO (1995) The design of multilevel multiharmonic signals for system identification. IEE Proc Control Theory Appl 142:247–252

Newman DJ (1965) An L1 extremal problem for polynomials. Proc Am Math Soc 16:1287–1290

Oliva Uribe D, Schoukens J, Stroop R (2018) Improved tactile resonance sensor for robotic assisted surgery. Mech Syst Sig Process 99:600–610

Pintelon R, Schoukens J (2012) System identification: a frequency domain approach. Wiley, Hoboken, NJ

Pintelon R, Louarroudi E, Lataire J (2014) Quantifying the time-variation in FRF measurements using random phase multisines with nonuniformly spaced harmonics. IEEE Trans Instrum Meas 63:1384–1394

Roinila T, Vilkko M, Sun J (2014) Online grid impedance measurement using discrete-interval binary sequence injection. IEEE J Emerg Sel Top Power Electron 2:985–993

Rudin W (1959) Some theorems on Fourier coefficients. Proc Am Math Soc 10:855–859

Sanchez B, Louarroudi E, Jorge E, Cinca J, Bragos R, Pintelon R (2013) A new measuring and identification approach for time-varying bioimpedance using multisine electrical impedance spectroscopy. Physiol Meas 34:339–357

Schoukens J, Pintelon R, Rolain Y, Dobrowiecki T (2001) Frequency response function measurements in the presence of nonlinear distortions. Automatica 37:939–946

Schroeder MR (1970) Synthesis of low-peak-factor signals and binary sequences with low autocorrelation. IEEE Trans Inf Theory 16:85–89
Shapiro HS (1951) Extremal problems for polynomials. M.S. thesis, Massachusetts Institute of Technology, MA
Stoev J, Schoukens J (2016) Nonlinear system identification—application for industrial hydro-static drive-line. Control Eng Pract 54:154–165
Tan AH, Godfrey KR (2004) An improved routine for designing multi-level multi-harmonic signals. In: Proceedings of the UKACC international conference on control (paper ID–027), Bath, UK, 6–9 Sept
Tan AH, Godfrey KR, Barker HA (2005) Design of computer-optimized pseudo-random maximum length signals for linear identification in the presence of nonlinear distortions. IEEE Trans Instrum Meas 54:2513–2519
van den Bos A, Krol RG (1979) Synthesis of discrete-interval binary signals with specified Fourier amplitude spectra. Int J Control 30:871–884
van der Maas R, van der Maas A, Dries J, de Jager B (2016) Efficient nonparametric identification for high-precision motion systems: a practical comparison based on a medical X-ray system. Control Eng Pract 56:75–85
Van der Ouderaa E, Schoukens J, Renneboog J (1988) Peak factor minimization using a time-frequency swapping algorithm. IEEE Trans Instrum Meas 37:145–147
Wahlberg B, Hjalmarsson H, Annergren M (2010) On optimal input design in system identification for control. In: Proceedings of the IEEE conference on decision and control, Atlanta, GA, 15–17 Dec, pp 5548–5553
Widanage WD, Barai A, Chouchelamane GH, Uddin K, McGordon A, Marco J, Jennings P (2016) Design and use of multisine signals for Li–ion battery equivalent circuit modelling. Part 1: signal design. J Power Sources 324:70–78
Yang Y, Wang L, Wang P, Yang X, Zhang F, Wen H, Teng Z (2015) Design of tri-level excitation signals for broadband bioimpedance spectroscopy. Physiol Meas 36:1995–2007

Chapter 4
Signal Design for Multi-input System Identification

4.1 Uncorrelated Design

For the identification of multi-input systems, methods using simultaneous perturbation of all the system inputs present significant benefits over those using sequential perturbation of the individual system inputs. An important advantage is that less time is spent waiting for the system to reach steady state. In addition to this, each transfer function is estimated with the process in the same conditions. This may not be true if sequential inputs are applied (Tan et al. 2015) since ambient conditions may differ between experiments. Identification of each of the subsystems is more straightforward if the system inputs comprise a set of signals that are uncorrelated with one another.

A set of signals is uncorrelated with one another if the crosscorrelation function between any pair of two inputs u_a and u_b satisfies

$$R_{u_a u_b}(n) = 0 \quad \forall n, \quad \text{provided}\, u_a \neq u_b. \tag{4.1}$$

In the frequency domain, the signals do not share any common excited (nonzero) harmonics. Such a set of signals allows the effects of individual inputs to be decoupled at the system output(s). The concept is shown in Fig. 4.1 for a linear system consisting of two inputs. The power at a particular harmonic in the output is attributed to the input with power at that harmonic.

For many years, it was thought that it is not possible to derive uncorrelated periodic signals with the same period, but MacWilliams (1967) achieved a breakthrough by designing a class of pairs of uncorrelated binary signals of period $N = 2p^2$, where p is a prime. However, for the pair of MacWilliams signals with $N = 18$, the autocorrelation functions of both signals are far from being of delta-function form (Briggs and Godfrey 1976). Their frequency spectra are therefore far from being uniform, which greatly limits the use of the signals in system identification applications.

A. H. Tan and K. R. Godfrey, *Industrial Process Identification*, Advances in Industrial Control, https://doi.org/10.1007/978-3-030-03661-4_4

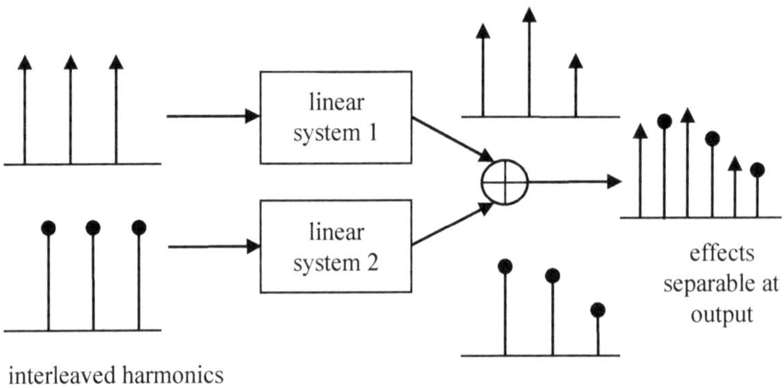

Fig. 4.1 Identification using uncorrelated signals

As such, for pseudorandom signal set design, focus has shifted to the design of signals with different periods (although they also share a common period). A lot more flexibility, including having signals in a set with the same period, can be achieved with multisine signals which have no constraint on the number of signal levels. In particular, there are two important methods for the design of sets of uncorrelated signals:

- modulation with rows of a Hadamard matrix
- design of a zippered spectrum

Note also that the methods to design a zippered spectrum are considerably different for multisine signal sets and for pseudorandom signal sets.

4.1.1 Modulation with Rows of a Hadamard Matrix

The Hadamard matrix is a square matrix of size $2^M \times 2^M$. It can be formed recursively using

$$H_{2^M} = \begin{bmatrix} H_{2^{M-1}} & H_{2^{M-1}} \\ H_{2^{M-1}} & -H_{2^{M-1}} \end{bmatrix}. \tag{4.2}$$

The first four Hadamard matrices in the series are

$$H_1 = [1], \tag{4.3}$$

$$H_2 = \begin{bmatrix} H_1 & H_1 \\ H_1 & -H_1 \end{bmatrix} = \begin{bmatrix} 1 & 1 \\ 1 & -1 \end{bmatrix}, \tag{4.4}$$

$$H_4 = \begin{bmatrix} H_2 & H_2 \\ H_2 & -H_2 \end{bmatrix} = \begin{bmatrix} 1 & 1 & 1 & 1 \\ 1 & -1 & 1 & -1 \\ 1 & 1 & -1 & -1 \\ 1 & -1 & -1 & 1 \end{bmatrix}, \tag{4.5}$$

$$H_8 = \begin{bmatrix} H_4 & H_4 \\ H_4 & -H_4 \end{bmatrix} = \begin{bmatrix} 1 & 1 & 1 & 1 & 1 & 1 & 1 & 1 \\ 1 & -1 & 1 & -1 & 1 & -1 & 1 & -1 \\ 1 & 1 & -1 & -1 & 1 & 1 & -1 & -1 \\ 1 & -1 & -1 & 1 & 1 & -1 & -1 & 1 \\ 1 & 1 & 1 & 1 & -1 & -1 & -1 & -1 \\ 1 & -1 & 1 & -1 & -1 & 1 & -1 & 1 \\ 1 & 1 & -1 & -1 & -1 & -1 & 1 & 1 \\ 1 & -1 & -1 & 1 & -1 & 1 & 1 & -1 \end{bmatrix}. \tag{4.6}$$

The Hadamard matrices can be generated using the function `hadamard` in MATLAB®. (MATLAB® is a registered product of The MathWorks, Inc.) To generate H_1, H_2, H_4 and H_8, type `H1=hadamard(1)`, `H2=hadamard(2)`, `H4=hadamard(4)` and `H8=hadamard(8)`, respectively, in the MATLAB command window.

The technique of modulation with rows of a Hadamard matrix was proposed by Briggs and Godfrey (1966). To obtain a set of M uncorrelated signals, a PRB signal is first generated (see Sects. 2.1 and 2.2 for the generation of PRB signals). Next, generate the Hadamard matrix $H_{2^{M-1}}$. Concatenate the PRB signal with itself 2^{M-1} times to produce a signal u_1. The signals u_1, u_2, u_3, ..., u_M in a set are obtained by modulating u_1 with rows $1, 2, 3, \ldots, 2^{M-2}+1$ of $H_{2^{M-1}}$, respectively. (The modulation with row 1 of $H_{2^{M-1}}$ leaves u_1 unchanged.) The signals are therefore obtained as follows:

- u_1: original signal concatenated 2^{M-1} times
- u_2: modulate with $[1\ -1]$
- u_3: modulate with $[1\ 1\ -1\ -1]$
- u_4: modulate with $[1\ 1\ 1\ 1\ -1\ -1\ -1\ -1]$
- u_5: modulate with $[1\ 1\ 1\ 1\ 1\ 1\ 1\ 1\ -1\ -1\ -1\ -1\ -1\ -1\ -1\ -1]$ and so on.

The signals in a set have a common period N corresponding to the period of u_M. The actual periods of u_1, u_2, u_3, ..., u_M are $N/2^{M-1}$, $N/2^{M-2}$, $N/2^{M-3}$, ..., N, respectively. In a common period N, u_1 therefore has 2^{M-1} subperiods, u_2 has 2^{M-2} subperiods, u_3 has 2^{M-3} subperiods and so on.

An example is shown here for the generation of four uncorrelated signals with a common period of $N = 24$. A QRB signal $[1\ -1\ 1]$ is first generated; this has a period of $N/2^{M-1} = 24/2^3 = 3$. The QRB signal is concatenated eight times with itself to produce $u_1 = [1\ -1\ 1\ 1\ -1\ 1\ 1\ -1\ 1\ 1\ -1\ 1\ 1\ -1\ 1\ 1\ -1\ 1\ 1\ -1\ 1\ 1\ -1\ 1]$. The generation of u_2, u_3 and u_4 is shown in Table 4.1.

Table 4.1 Generation of uncorrelated signal set through modulation with rows of a Hadamard matrix

i	$u_1(i)$	Row 2 (R2) of H_8	$u_2(i) =$ R2 × $u_1(i)$	Row 3 (R3) of H_8	$u_3(i) =$ R3 × $u_1(i)$	Row 5 (R5) of H_8	$u_4(i) =$ R5 × $u_1(i)$
1	1	1	1	1	1	1	1
2	−1	−1	1	1	−1	1	−1
3	1	1	1	−1	−1	1	1
4	1	−1	−1	−1	−1	1	1
5	−1	1	−1	1	−1	−1	1
6	1	−1	−1	1	1	−1	−1
7	1	1	1	−1	−1	−1	−1
8	−1	−1	1	−1	1	−1	1
9	1	1	1	1	1	1	1
10	1	−1	−1	1	1	1	1
11	−1	1	−1	−1	1	1	−1
12	1	−1	−1	−1	−1	1	1
13	1	1	1	1	1	−1	−1
14	−1	−1	1	1	−1	−1	1
15	1	1	1	−1	−1	−1	−1
16	1	−1	−1	−1	−1	−1	−1
17	−1	1	−1	1	−1	1	−1
18	1	−1	−1	1	1	1	1
19	1	1	1	−1	−1	1	1
20	−1	−1	1	−1	1	1	−1
21	1	1	1	1	1	−1	−1
22	1	−1	−1	1	1	−1	−1
23	−1	1	−1	−1	1	−1	1
24	1	−1	−1	−1	−1	−1	−1

The signal set can also be generated using MATLAB. To do this, type the following codes into the MATLAB command window:

```
%generate the Hadamard matrix
H8=hadamard(8);
%x is the original PRB signal
x=[1;-1;1];

%generate u1 by concatenating the signal x 8 times
u1=[x;x;x;x;x;x;x;x];

%generate u2
R2=H8(2,:);R2=R2';
R2=[R2;R2;R2];
u2=u1.*R2;
```

```
%generate u3
R3=H8(3,:);R3=R3';
R3=[R3;R3;R3];
u3=u1.*R3;

%generate u4
R5=H8(5,:);R5=R5';
R5=[R5;R5;R5];
u4=u1.*R5;
```

The DFT magnitudes of the signal set are shown in Fig. 4.2. Note that the DFT magnitudes are generally not uniform, except for u_1, u_2 and u_3 which have uniform DFT magnitudes at most of the excited harmonics with a small number of exceptions. For u_1, a smaller magnitude is observed at harmonic 0; for u_2, a smaller magnitude is observed at harmonic $N/2$; for u_3, this appears at harmonics $N/4$ and $3N/4$. In particular,

$$|U_1(k)| = \begin{cases} 2^{M-1} & k = 0 \\ 2^{M-1}\sqrt{\frac{N}{2^{M-1}}+1} & k \in \{2^{M-1}p | p \in \mathrm{Z}\}, k \neq 0 \\ 0 & \text{otherwise} \end{cases}, \tag{4.7}$$

$$|U_2(k)| = \begin{cases} 2^{M-1} & k = N/2 \\ 2^{M-1}\sqrt{\frac{N}{2^{M-1}}+1} & k \in \{2^{M-2}p | p \text{ odd}\}, k \notin N/2 \\ 0 & \text{otherwise} \end{cases}, \tag{4.8}$$

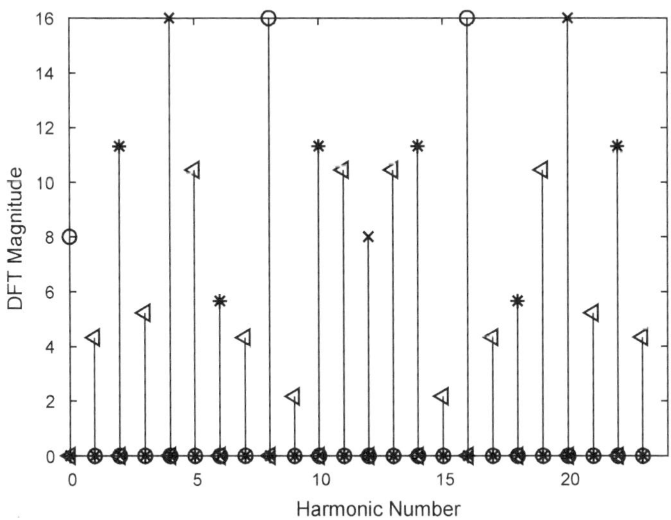

Fig. 4.2 DFT magnitudes of u_1 (circles), u_2 (crosses), u_3 (asterisks) and u_4 (triangles) using modulation with rows of a Hadamard matrix

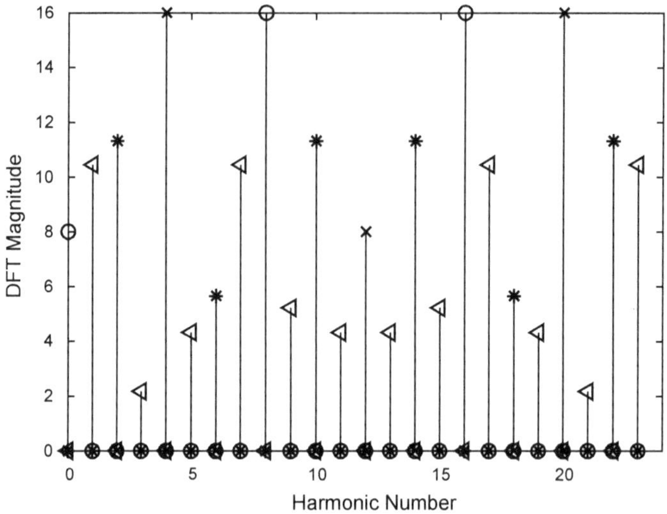

Fig. 4.3 DFT magnitudes of u_1 (circles), u_2 (crosses), u_3 (asterisks) and u_4 (triangles) using a variant of the modulation with rows of a Hadamard matrix

which in the case of $M = 2$, the signal set simply degenerates into a PRB signal (concatenated twice) and its corresponding inverse-repeat signal (see Sects. 2.1 and 2.2).

Roinila et al. (2013, 2018) and Jin et al. (2013, 2014) applied Hadamard modulation to obtain a set of uncorrelated signals. In particular, Jin et al. (2013, 2014) referred to these signals as multidimensional inverse M sequences, when the original sequence is an MLB sequence. Working with signal levels 0 and 1, they proposed an M sequence generator consisting of frequency dividers, each of which performs divide-by-two operation. For example, a [0 1 0 1 0 1 0 1] sequence after passing through a frequency divider gives a sequence [0 0 1 1 0 0 1 1] which has half the frequency of the original sequence. After passing through another frequency divider, the resulting sequence is [0 0 0 0 1 1 1 1] which has now a quarter of the frequency of the original sequence. In hardware, divide-by-two frequency dividers can be easily implemented using D flip-flops. Interestingly, Jin et al. (2014) applied these signals in closed loop. While these signals ensure identifiability of the system, the inputs to the process are no longer uncorrelated with one another due to the feedback.

It is also possible to obtain uncorrelated signal sets by modulating with some other rows of the Hadamard matrix. In particular, the signals $u_1, u_2, u_3, \ldots, u_M$ in a set may alternatively be obtained by modulating u_1 with rows 1, 2, 4, ..., 2^{M-1} of $H_{2^{M-1}}$. The DFT magnitudes of a set of four uncorrelated signals with a common period of $N = 24$ for this variant of the Hadamard modulation are plotted in Fig. 4.3. Note that u_1, u_2 and u_3 are the same as before (compare with Fig. 4.2).

4.1.2 Design of a Zippered Spectrum Using Multisine Signals

If the system can accept signals with a large number of levels, the zippered spectrum can be achieved easily using a set of uncorrelated multisines. The excited harmonics are chosen such that there are no common excited harmonics between any signals in a set. The design of a zippered spectrum with 'snowing' is described in Rivera et al. (2009). Harmonic suppression can be incorporated easily. The suppression of even harmonics ensures that the effects of even order nonlinearities can be separated and completely eliminated at the system output. Further suppression of harmonic multiples of three can also be considered for reducing the effects of odd order nonlinearities on the linear estimates. See Sect. 5.4.2 for an application example on a mist reactor.

An adaptive procedure was proposed by Martín et al. (2016) which refines the perturbation signal parameters during experimental execution. The rationale behind this technique is that typically, only limited information is available before the start of the identification test and hence, the perturbation signal may not be very well designed. As information gradually becomes available during the experiment, the amplitude, number of signals periods as well as frequency content of the signal set can be manipulated while the experiment is running so that the identification test can be completed in the shortest possible duration subject to achieving a minimum accuracy in terms of the parameter estimates. In particular, the adjustment in the frequency content is done by halving the number of excited harmonics in each signal in a set, while at the same time increasing the power of the excited harmonics, thus maintaining the same input power. The concept is illustrated in Fig. 4.4 for a system with three inputs. In Fig. 4.4, the number of excited harmonics is halved from the top plot to the middle one, and halved again from the middle plot to the bottom one.

4.1.3 Design of a Zippered Spectrum Using Pseudorandom Signals

The design is more complicated when the system can accept signals with a limited number of signal levels. A search can be made for such signal sets. There is also a software package available known as the Input-Signal-Creator which is based on the work in Barker et al. (2014) (see Sect. 8.5). To ease subsequent explanation, three classes of input signals are defined based on their spectra across their individual periods (Tan et al. 2015).

- Class 0 signals: the power $|U(k)|^2$ of the kth harmonic of the signal is constant for all $0<k<N/2$,
- Class 1 signals: as Class 0 signals except that $|U(k)|^2=0$ when k is a multiple of 2;
- Class 2 signals: as Class 1 signals except that $|U(k)|^2=0$ also when k is a multiple of 3.

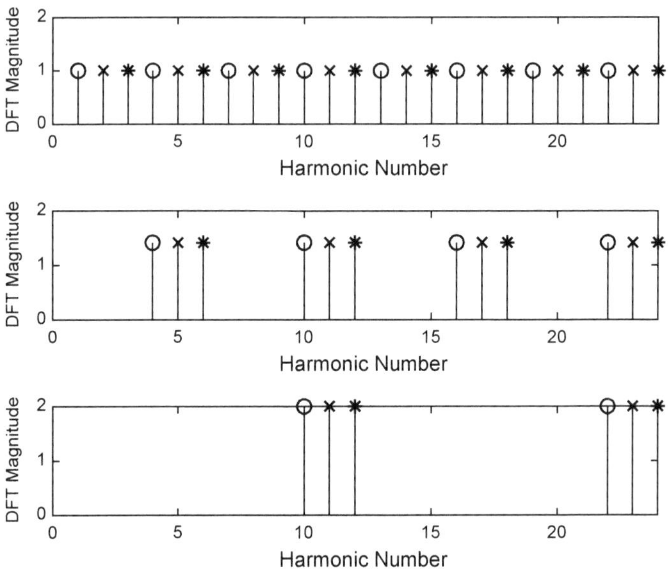

Fig. 4.4 Changes in input frequency content in the adaptive design of zippered spectrum using multisine signals. Two harmonic reduction steps are shown going from the top plot to the middle one and from the middle plot to the bottom one. The excited harmonics for inputs u_1, u_2 and u_3 are shown using circles, crosses and asterisks, respectively

A possible method for creating a set of M input signals uses a corresponding set of M signals $u_1, u_2, u_3, \ldots, u_M$ with periods N, $N/2$, $N/4$, …, $N/2^{M-1}$, as detailed in Tan et al. (2015). (Note the notational difference with respect to Sect. 4.1.1; this is for consistency with literature.) To apply this technique, 2^{M-1} must be a divisor of the common period N. Let the first input signal $u_1(i)_N = u(i)_N$. The second input signal $u_2(i)_N$ is obtained by concatenating (or aggregating) two periods of the signal $u(i)_{N/2}$; the latter can be viewed as a source signal where the former originates from. The third input signal $u_3(i)_N$ is obtained by concatenating four periods of the source signal $u(i)_{N/4}$. A similar procedure is applied up to the Mth input signal $u_M(i)_N$; this is obtained by concatenating 2^{M-1} periods of a signal with period $N/2^{M-1}$. The signal set is given by

$$
\begin{aligned}
u_1(i)_N &= u(i)_N \\
u_2(i)_N &= u(i)_{N/2}u(i)_{N/2} \\
u_3(i)_N &= u(i)_{N/4}u(i)_{N/4}u(i)_{N/4}u(i)_{N/4} \\
&\ldots \\
u_M(i)_N &= u(i)_{N/2^{M-1}}u(i)_{N/2^{M-1}} \ldots u(i)_{N/2^{M-1}}.
\end{aligned}
\tag{4.9}
$$

The excited harmonics in the signal set depend on the signal class of the source signals. The signals have uniform DFT magnitudes at the excited harmonics in the range $0<k<N/2$. The main difference of this technique with the technique of modulation with rows of a Hadamard matrix is that here, the signals generally do not all originate from a common single source signal.

When all the input signals in Eq. 4.9 are created from Class 1 signals, the excited harmonics are given by

$$\begin{aligned}
&|U_1(k)|_N^2 \neq 0 \ \text{ when } k = 1, 3, 5, 7, 9, 11, \ldots \\
&|U_2(k)|_N^2 \neq 0 \ \text{ when } k = 2, 6, 10, 14, 18, 22, \ldots \\
&|U_3(k)|_N^2 \neq 0 \ \text{ when } k = 4, 12, 20, 28, 36, 44, \ldots \\
&\ldots \\
&|U_M(k)|_N^2 \neq 0 \ \text{ when } k = 2^{M-1}(1, 3, 5, 7, 9, 11, \ldots)
\end{aligned} \tag{4.10}$$

from which it is clear that the signals do not share any common excited harmonics and so they are spectrally independent and uncorrelated. However, a disadvantage of the method is that the frequency resolution decreases at 2^{M-1}, as the spacing between the excited harmonics increases by a factor of two going from the first input to the second, from the second input to the third, and so on. In order to achieve a given frequency resolution, the measurement time will need to be increased proportionally.

When all the input signals in Eq. 4.9 are created from Class 2 signals, the excited harmonics are given by

$$\begin{aligned}
&|U_1(k)|_N^2 \neq 0 \ \text{ when } k = 1, 5, 7, 11, 13, 17, \ldots \\
&|U_2(k)|_N^2 \neq 0 \ \text{ when } k = 2, 10, 14, 22, 26, 34, \ldots \\
&|U_3(k)|_N^2 \neq 0 \ \text{ when } k = 4, 20, 28, 44, 52, 68, \ldots \\
&\ldots \\
&|U_M(k)|_N^2 \neq 0 \ \text{ when } k = 2^{M-1}(1, 5, 7, 11, 13, 17, \ldots)
\end{aligned} \tag{4.11}$$

from which it can be seen that the signals do not share any common excited harmonics. Again, the frequency resolution decreases at 2^{M-1}.

As Class 0 signals do not incorporate harmonic suppression, their use is more restricted. However, the method in Eq. 4.9 can still be applied when a pair of signals (two signals) in a set is required. In this case, the first input signal $u_1(i)_N$ is a Class 1 signal and the second input signal $u_2(i)_N$ is obtained by concatenating two periods of a Class 0 signal $u(i)_{N/2}$. The excited harmonics are given by

$$\begin{aligned}
&|U_1(k)|_N^2 \neq 0 \ \text{ when } k = 1, 3, 5, 7, 9, 11, \ldots \\
&|U_2(k)|_N^2 \neq 0 \ \text{ when } k = 2, 4, 6, 8, 10, 12, \ldots
\end{aligned} \tag{4.12}$$

from which it can be observed that the signals do not share any common excited harmonics and so they are spectrally independent and uncorrelated. If the signals are a PRB signal and its corresponding inverse-repeat version, this method overlaps with the technique of modulation with rows of a Hadamard matrix.

Other combinations which do not follow Eq. 4.9 are also possible, although they are typically found through a careful search. Some examples using PRML signals, including truncated PRML signals, are given in Tan et al. (2009). (For the design of the signals, refer to Sect. 2.3.) In particular, two of these examples for a three-input set will be illustrated here. The design capitalises on the fact that if the primitive signals are uncorrelated with one another, then the resulting PRML signals are also uncorrelated with one another. The relationship is described by Eq. 2.22.

The first example is based on GF(7) with $n = 2$, where the set of uncorrelated primitive signals were found through a search procedure. Those corresponding to the three inputs are

- u_1: [1 1 0 −1 −1 0],
- u_2: [1 −1 0 1 −1 0],
- u_3: [1 −1 1 −1 1 −1].

The DFT magnitudes of the signal set are plotted in Fig. 4.5 using signals with maximum and minimum amplitudes of 1 and −1, respectively. The signal amplitudes can be adjusted in practice, if necessary.

The second example is based on GF(11) with $n = 2$, where the primitive signals corresponding to the three inputs are:

- u_1: [1 −1 −1 0 −1 −1 1 1 0 1],
- u_2: [1 −1 −1 1 0 1 −1 −1 1 0],
- u_3: [1 −1 1 −1 1 −1 1 −1 1 −1].

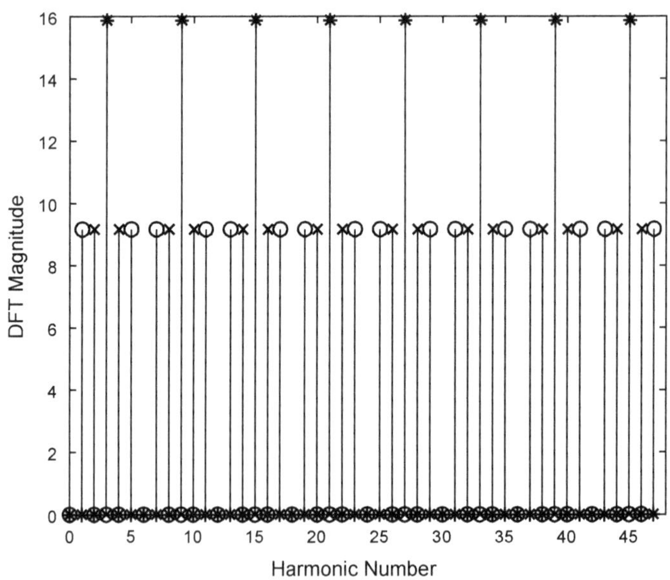

Fig. 4.5 DFT magnitudes of u_1 (circles), u_2 (crosses) and u_3 (asterisks) for an uncorrelated signal set from GF(7)

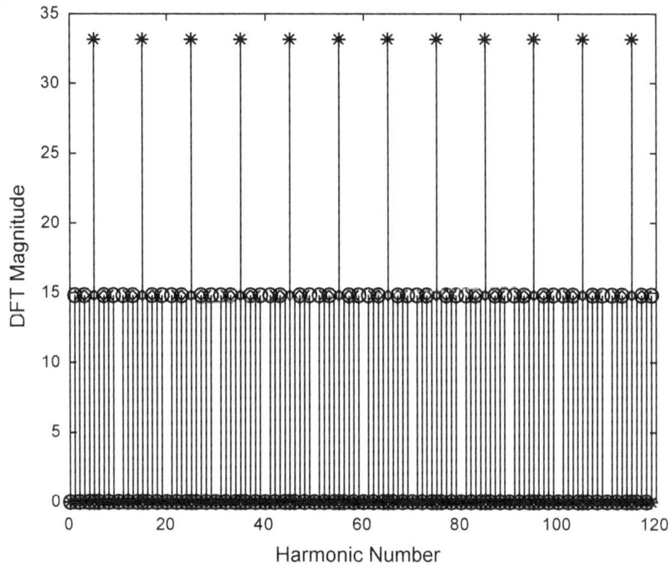

Fig. 4.6 DFT magnitudes of u_1 (circles), u_2 (crosses) and u_3 (asterisks) for an uncorrelated signal set from GF(11)

The DFT magnitudes of the signal set are plotted in Fig. 4.6.

4.1.4 Application Example

The system to be identified is a two-input two-output system comprising a cascade of two thermoelectric modules which form part of a microfluidic platform for genetic basis of disease diagnosis (Kaigala et al. 2010). Details of the simulation results are presented in Tan et al. (2015). The physical layout consists of a chip layer, two thermoelectric modules in a cascade arrangement and a heat sink. High purity copper plates are placed between each of the four layers for enhancing heat transfer, maintaining temperature uniformity and serving as a physical housing for the thermal sensors.

The two-input two-output system can be modelled by $\mathbf{Y} = \mathbf{G}_{\text{actual}}\mathbf{U}$, where $\mathbf{Y}$ and $\mathbf{U}$ are the vectors of the Laplace transform of $\mathbf{y} = [\,y_1\ y_2]^{\text{T}}$ and $\mathbf{u} = [\,u_1\ u_2]^{\text{T}}$, respectively. The inputs u_1 and u_2 denote the voltage input of the upper and lower thermoelectric modules, respectively. The output y_1 represents the temperature difference between the upper and middle copper plates whereas the output y_2 represents the temperature difference between the middle and lower copper plates. Kaigala et al. (2010) applied a set of random binary signals of length 4000 to the inputs as these signals possess low crest factor. A sampling interval of 1 s was selected. The identification was carried out over the frequency range between 0.01 and 0.05 of the Nyquist frequency using the System Identification Toolbox in MATLAB (Ljung and Singh 2012). As the system parameters vary across the operating range of interest, three local models were estimated at three different temperature regions. At the middle region corresponding to 72 °C, the estimated transfer function matrix was

$$\mathbf{G}(s)_{\text{actual}} = \begin{bmatrix} \frac{0.03438}{17.35s+1} & \frac{-1.031s-0.001137}{563.2s^2+68.57s+1}\mathrm{e}^{-1.34s} \\ \frac{-0.7401s+0.001507}{488.4s^2+65.61s+1}\mathrm{e}^{-0.757s} & \frac{0.03537}{27.05s+1} \end{bmatrix} \tag{4.13}$$

which subsequently served as the baseline actual transfer function matrix.

Estimates of the FRFs of the subsystems $G_{12}(s)$ and $G_{21}(s)$ are required in the frequency range between $f_{\text{min}} = 0.00055$ Hz and $f_{\text{max}} = 0.5$ Hz. The sampling interval t_s needs to satisfy

$$t_s \leq 1/(2f_{\text{max}}). \tag{4.14}$$

The sampling interval was set to $t_s = 1$ s following the choice in Kaigala et al. (2010). For selecting the signal period, the guideline in Tan et al. (2015) given by

$$N/2^i = 1/(t_s \times f_{\text{resolution}}) \geq 2f_{\text{max}}/f_{\text{min}}; \quad i = 0, 1, \ldots, M-1 \tag{4.15}$$

was applied; this dictates that $N \geq 3636$.

The objective of the experiments was to show that this system could be satisfactorily identified using sets of uncorrelated signals in simultaneous perturbation. Two sets of signals were used. For the first signal set (Signal Set A), a Class 1/Class 0 signal pair following Eq. 4.12 was utilised in which the signal u_1 was a Class 1 binary signal with the common period 3814 whereas the signal u_2 was a Class 0 binary signal with period 1907. These were a QRB signal of period 1907 and its inverse-repeat pair. Signal u_1 has excited harmonics at 1, 3, 5, 7, 9, 11, … whereas signal u_2 has excited harmonics at 2, 4, 6, 8, 10, 12, … across the common period. For the second signal set (Signal Set B), Class 2 signals were designed according to Eq. 4.11. The signal u_1 was a Class 2 ternary signal with the common period 3792 whereas the signal u_2 was a Class 2 ternary signal with period 1896. In particular, u_1 was a truncated PRML signal from GF(631), with $n = 2$ and 105 subperiods in a period of the primitive signal while u_2 was a truncated PRML signal from GF(157), with $n = 2$ and 13 subperiods in a period of the primitive signal. Signal u_1 has excited harmonics at 1, 5, 7, 11, 13, 17, … whereas signal u_2 has excited harmonics at 2, 10, 14, 22, 26, 34, … across the common period. Some important properties of the signal sets are summarised in Table 4.2.

Only one steady-state period was utilised for estimation. The transfer functions were estimated using the maximum likelihood estimator implemented in the Estimator for Linear Systems (ELiS) in the Frequency Domain System Identification Toolbox in MATLAB (Kollár 1994). In Tan et al. (2015), the data used in the identification were limited to the range between the frequency of the smallest harmonic in

Table 4.2 Properties of uncorrelated signal sets used

Signal set	Signal	Class	Number of levels	PIPS (%)
A	u_1	1	2	100
	u_2	0	2	100
B	u_1	2	3	81.59
	u_2	2	3	81.39

the input signals and 0.05 of the Nyquist frequency, where the upper limit was chosen to follow that used by Kaigala et al. (2010). Restricting the identification to this frequency range ensured that sufficient emphasis was given to the excited harmonics close to the resonance and minimised the effects of noise and nonlinearities which were more significant at higher frequencies.

Three tests were carried out under different scenarios. These are described in Tan et al. (2015). The tests were

- Test 1: The system was corrupted by additive noise in the outputs. The amount of noise was selected to result in an SNR of 20 dB for each output (y_1 and y_2) for Signal Set A. This was achieved using noise with RMS values of 0.137 and 0.111 for y_1 and y_2, respectively. The same amount of noise was then applied also to Signal Set B for fair comparison.
- Test 2: The system was corrupted by even-order nonlinearity in the form of quadratic nonlinearity. The amount of nonlinearity was selected to give a signal-to-nonlinearity ratio, $20\log_{10}\left(\frac{\text{RMS(signal)}}{\text{RMS(nonlinearity)}}\right)$, of 20 dB for each output (y_1 and y_2) for Signal Set A. This was achieved by the nonlinear functions $f(y_1) = y_1 + 0.043y_1^2$ and $f(y_2) = y_2 + 0.048y_2^2$ for y_1 and y_2, respectively. The same nonlinear functions were then applied also to Signal Set B for fair comparison.
- Test 3: The system was corrupted by odd-order nonlinearity in the form of cubic nonlinearity. The amount of nonlinearity was selected to give a signal-to-nonlinearity ratio of 20 dB for each output (y_1 and y_2) for Signal Set A. This was achieved by the nonlinear functions $f(y_1) = y_1 + 0.015y_1^3$ and $f(y_2) = y_2 + 0.018y_2^3$ for y_1 and y_2, respectively. The same nonlinear functions were then applied also to Signal Set B for fair comparison.

The Nyquist plots corresponding to the estimated FRFs for Test 1 and Test 3 are depicted in Figs. 4.7 and 4.8, respectively. For Test 1 (Fig. 4.7), the Nyquist plots for both Signal Set A and Signal Set B match the actual ones very well, showing that the signals are very robust in the presence of noise. The plots of Test 2 are quite similar to those for Test 1, with only a minor deterioration in the model accuracy being observed. These plots are therefore not shown here, but it is noted that the signals are robust also to the presence of even order nonlinearity. However, a bias is clearly present in Fig. 4.8; this can be attributed to the systematic effect of the cubic nonlinearity.

The parameters of the estimated transfer function matrix as given in Tan et al. (2015) were as follows:

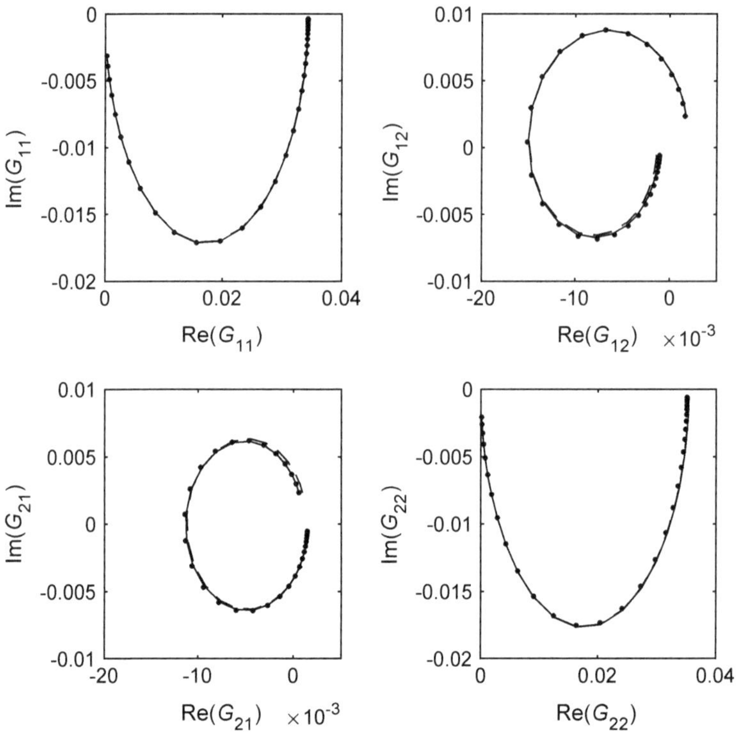

Fig. 4.7 Nyquist plots for test 1. Solid line: actual; dashed line: Set A; dotted line: Set B

Test1:

$$\mathbf{G}_{\text{setA}} = \begin{bmatrix} \frac{0.03430}{17.35s+1} & \frac{-0.9832s-0.001372}{539.7s^2+65.60s+1}e^{-1.36s} \\ \frac{-0.7329s+0.001503}{470.2s^2+65.53s+1}e^{-1.02s} & \frac{0.03531}{27.06s+1} \end{bmatrix}, \tag{4.16}$$

$$\mathbf{G}_{\text{setB}} = \begin{bmatrix} \frac{0.03436}{17.34s+1} & \frac{-1.007s-0.001027}{554.2s^2+66.47s+1}e^{-1.27s} \\ \frac{-0.7486s+0.001485}{488.2s^2+65.66s+1}e^{-0.731s} & \frac{0.03510}{26.97s+1} \end{bmatrix}. \tag{4.17}$$

Test 2:

$$\mathbf{G}_{\text{setA}} = \begin{bmatrix} \frac{0.03376}{16.99s+1} & \frac{-1.070s-0.001569}{587.3s^2+72.68s+1}e^{-1.48s} \\ \frac{-0.7174s+0.001893}{466.9s^2+65.15s+1}e^{-1.11s} & \frac{0.03520}{26.94s+1} \end{bmatrix}, \tag{4.18}$$

$$\mathbf{G}_{\text{setB}} = \begin{bmatrix} \frac{0.03452}{17.36s+1} & \frac{-0.9521s-0.001574}{540.1s^2+61.85s+1}e^{-1.24s} \\ \frac{-0.7409s+0.001696}{508.6s^2+65.30s+1}e^{-0.599s} & \frac{0.03542}{27.15s+1} \end{bmatrix}. \tag{4.19}$$

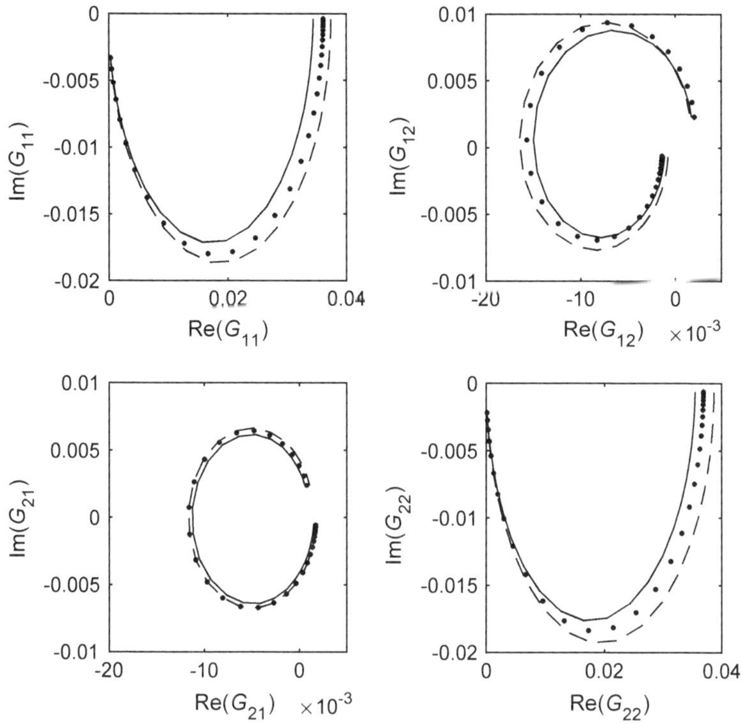

Fig. 4.8 Nyquist plots for Test 3. Solid line: actual; dashed line: Set A; dotted line: Set B

Test 3:

$$\mathbf{G}_{\text{setA}} = \begin{bmatrix} \frac{0.03739}{17.46s+1} & \frac{-1.121s-0.0007619}{592.2s^2+67.98s+1}e^{-1.13s} \\ \frac{-0.7826s+0.001770}{470.6s^2+66.99s+1}e^{-1.05s} & \frac{0.03868}{27.35s+1} \end{bmatrix}, \quad (4.20)$$

$$\mathbf{G}_{\text{setB}} = \begin{bmatrix} \frac{0.03608}{17.22s+1} & \frac{-1.072s-0.001417}{542.9s^2+68.10s+1}e^{-1.47s} \\ \frac{-0.7861s+0.001745}{489.9s^2+67.35s+1}e^{-0.846s} & \frac{0.03683}{26.82s+1} \end{bmatrix}. \quad (4.21)$$

The percentage mean magnitudes of the complex error in the frequency response defined by $\frac{E[|\mathbf{G}_{\text{actual}}(j\omega)-\mathbf{G}_{\text{estimated}}(j\omega)|]}{E[|\mathbf{G}_{\text{actual}}(j\omega)|]} \times 100\%$ were computed and the values are given in Table 4.3.

From Figs. 4.7 and 4.8 as well as Table 4.3, the following observations can be made:

- In Test 1, Signal Set A achieved a more accurate estimation overall. This can be attributed to the fact that Signal Set A has a larger power within amplitude constraints since the signals are binary.

Table 4.3 Percentage mean magnitudes of the complex error $\frac{E[|G_{actual}(j\omega)-G_{estimated}(j\omega)|]}{E[|G_{actual}(j\omega)|]} \times 100\%$ in the frequency response

Test	Signal set	Error in G_{11} (%)	Error in G_{12} (%)	Error in G_{21} (%)	Error in G_{22} (%)
1	A	0.233	0.694	0.986	0.191
	B	0.041	0.980	1.122	0.611
2	A	1.294	3.185	2.506	0.305
	B	0.381	2.420	1.234	0.192
3	A	8.442	8.190	4.947	8.701
	B	5.318	5.497	4.119	4.633

- In Test 2, Signal Set B resulted in more accurate estimates. Both signals in Signal Set B originate from source signals which have even harmonics suppressed, compared to only one of the signals in Signal Set A. In addition to this, the larger power of Signal Set A led to higher excitation of the nonlinearity.
- In Test 3, Signal Set B again resulted in more accurate estimates. However, the bias in the estimation led to error values that were significantly larger compared with those in Test 2.

This example demonstrates the feasibility of applying simultaneous perturbation to a system with multiple inputs. In comparison with sequential perturbation, the simultaneous perturbation approach reduces the experimentation time since transient removal is only required once compared to twice (once for each input) in the case of sequential perturbation. If a single period N is needed for transient removal, sequential perturbation necessitates a measurement time of approximately $4N$. In contrast, the simultaneous perturbation approach requires approximately $3N$; this corresponds to a time saving of roughly 25%.

4.2 Phase-Shifting Design

The phase-shifting design, described by Briggs and Godfrey (1966) as the input phase separation method, uses appropriately delayed versions of a PRB signal as the system inputs. (Although other classes of signals can also be applied, typically a PRB signal is applied.) The measurement period must be sufficiently long so that the output responses due to different input phases are separable in the input–output crosscorrelation function.

An example is used to illustrate the concept. Referring to Fig. 4.9, let u_1 be a PRB signal u of period N. The input u_2 is the same PRB signal shifted by, say, $\tau_1 \approx N/3$. This shift can be adjusted if more information about the system is available. The input u_3 is the same PRB signal shifted by, say, $\tau_2 \approx 2N/3$ (again, this can be

Fig. 4.9 Block diagram of a 3-input system

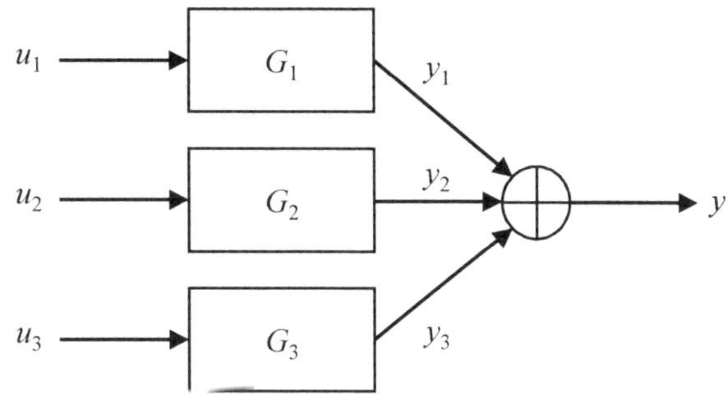

adjusted if necessary). The transfer functions are denoted by G_1, G_2 and G_3. The crosscorrelation function

$$R_{uy}(n) = \frac{1}{N}\sum_{i=1}^{N} u(i)y(i+n)$$
$$= \frac{1}{N}\sum_{i=1}^{N} u(i)[y_1(i+n) + y_2(i+n) + y_3(i+n)] \tag{4.22}$$

gives

$$R_{uy}(n) = R_{uy_1}(n) + R_{uy_2}(n) + R_{uy_3}(n). \tag{4.23}$$

The reference starting point for $R_{uy_1}(n)$ is $n = 0$. However, due to the phase shifts between u_1, u_2 and u_3, the reference starting points for $R_{uy_2}(n)$ and $R_{uy_3}(n)$ are τ_1 and τ_2, respectively. Provided N is sufficiently large, the effects of the individual inputs can be separated at the system output.

Let the system in Fig. 4.9 be described by $G_1(z^{-1}) = \frac{1}{1-0.9z^{-1}}$, $G_2(z^{-1}) = \frac{1}{1-0.7z^{-1}}$ and $G_3(z^{-1}) = \frac{1}{1-1.5z^{-1}+0.7z^{-2}}$. Using an MLB signal with $N = 127$ as u_1, the same signal is shifted by 42 bits to form u_2 and shifted by 84 bits to form u_3. The input signals are plotted in Fig. 4.10. The input-output crosscorrelation function is plotted in Fig. 4.11, where it can be seen that the three responses due to the three different transfer functions are separable.

4.3 Identification of Ill-Conditioned Processes

4.3.1 Problem of Ill-Conditioning

Consider a multivariable system with n inputs and n outputs given by

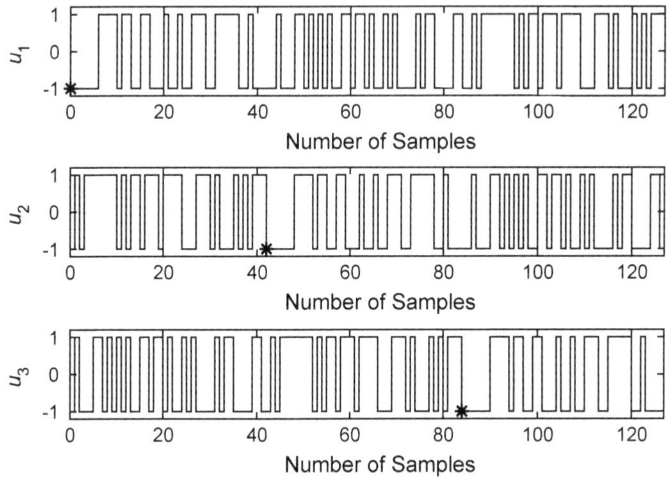

Fig. 4.10 Input signals using phase-shifting design. The asterisk in each plot marks the reference point

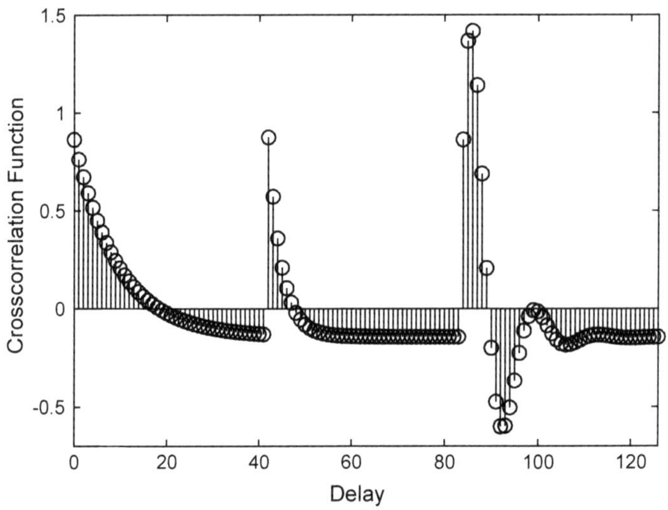

Fig. 4.11 Normalised input-output crosscorrelation function using phase-shifting design

$$\mathbf{Y}(s) = \mathbf{G}(s)\mathbf{U}(s), \mathbf{G}(s) = \begin{bmatrix} G_{11}(s) & G_{12}(s) & \ldots & G_{1n}(s) \\ G_{21}(s) & G_{22}(s) & \ldots & G_{2n}(s) \\ \vdots & \vdots & & \vdots \\ G_{n1}(s) & G_{n2}(s) & \ldots & G_{nn}(s) \end{bmatrix}, \tag{4.24}$$

where $\mathbf{U}(s) = \begin{bmatrix} U_1(s) \; U_2(s) \; \ldots \; U_n(s) \end{bmatrix}^T$ and $\mathbf{Y}(s) = \begin{bmatrix} Y_1(s) \; Y_2(s) \; \ldots \; Y_n(s) \end{bmatrix}^T$ are vectors of the Laplace transform of the inputs and outputs, respectively, and $\mathbf{G}(s)$ is the transfer function matrix. It is assumed that at least one element in each row and column of $\mathbf{G}(s)$ is nonzero. Applying singular value decomposition

$$\mathbf{G}(s) = \begin{bmatrix} \mathbf{b}_1(s) \, \mathbf{b}_2(s) \ldots \mathbf{b}_n(s) \end{bmatrix} \begin{bmatrix} \sigma_1(s) & 0 & \ldots & 0 \\ 0 & \sigma_2(s) & \ldots & 0 \\ \vdots & \vdots & & \vdots \\ 0 & 0 & \ldots & \sigma_n(s) \end{bmatrix} \begin{bmatrix} \mathbf{a}_1^T(s) \\ \mathbf{a}_2^T(s) \\ \vdots \\ \mathbf{a}_n^T(s) \end{bmatrix}, \tag{4.25}$$

where $\mathbf{a}_i(s)$ and $\mathbf{b}_i(s)$ are the singular vectors of the input and output, respectively. These singular vectors form orthonormal basis sets. The singular values are arranged such that $\sigma_1(s) \geq \sigma_2(s) \geq \cdots \geq \sigma_n(s)$. Due to the orthonormality, the process $\mathbf{G}(s)$ will transform the ith input singular vector into the ith output singular vector scaled by a gain of σ_i.

The system is ill-conditioned at a particular complex frequency $s = j\omega$, where ω is the angular frequency, if the condition number $\gamma(s) = \sigma_1(s)/\sigma_n(s) >> 1$. For an ill-conditioned process, the input vectors are amplified very differently depending on their directions. This leads to high (or strong) and low (or weak) gain directions which refer to singular vectors corresponding to the maximum and minimum singular values, respectively. Some examples of systems which are ill-conditioned are a distillation column (Waller and Böling 2005), a fluid catalytic cracking system (Bruwer and MacGregor 2006) and a gasifier (Chin and Munro 2003). If the system is estimated using an excitation of equal magnitude in all input directions, the response in the low gain direction will be very small compared with those in the other directions. This may subsequently result in an incorrect estimate of the system determinant (Waller and Böling 2005). While the largest singular value can be easily identified, the identification of the smallest singular value presents a challenge. Unfortunately, for successful application of model-based control, accurate identification of the smallest singular value, besides the other singular values, is very important (Li and Lee 1996; Rasmussen and Jørgensen 1999). This motivates research into the design of perturbation signals which specifically handle the issue of ill-conditioning.

One of the methods to identify ill-conditioned processes is through the use of rotated inputs (Conner and Seborg 2004), where some of the inputs are 'rotated' at certain angles which are either assumed to be approximately known a priori or obtained by trial and error. An adaptive design of experiments for rotated inputs can be applied for estimating the model order (Misra and Nikolaou 2017). The approach proposed by Li and Lee (1996) utilises a combination of open-loop- and closed-loop identification. In Zhu and Stec (2006), low amplitude uncorrelated binary signals were used to perturb the high gain direction while high amplitude correlated binary signals were utilised to excite the low gain direction. The uncorrelated and correlated binary signals could be applied alternately, or simultaneously, in an additive manner. The technique can be applied in both open loop and closed loop. The idea was extended to the frequency domain with the modified zippered spectrum design (Rivera et al. 2009) consisting of alternating pattern of correlated harmonics with high levels of power and uncorrelated harmonics with lower levels of power. The method is highly flexible in terms of signal period and frequency content. It is also possible to subject the design to criteria related to plant-friendliness. More recently, an open-loop method based on virtual transfer function between inputs (VTFBI) was introduced (Tan and Yap 2012). This technique will be discussed in greater detail in Sect. 4.3.2.

4.3.2 Virtual Transfer Function Between Inputs

The VTFBI design for 2×2 systems is first described. Extension to higher dimensional systems is treated later in the section.

Define each input $u_i(t)$, $i = \{1, 2\}$, as comprising of a single correlated harmonic $p_i(t)$ and the sum of several uncorrelated harmonics $q_i(t)$ such that

$$u_1(t) = p_1(t) + q_1(t), \tag{4.26}$$

$$u_2(t) = p_2(t) + q_2(t), \tag{4.27}$$

where

$$p_1(t) = A_{p_1} \sin\left(\omega_0 t + \phi_{p_1}\right), \tag{4.28}$$

$$p_2(t) = A_{p_2} \sin\left(\omega_0 t + \phi_{p_2}\right), \tag{4.29}$$

$$q_1(t) = \sum_{l=1}^{N_l} A_{k_1(l)} \sin\left(\omega_{k_1(l)} t + \phi_{k_1(l)}\right), \tag{4.30}$$

$$q_2(t) = \sum_{m=1}^{N_m} A_{k_2(m)} \sin\left(\omega_{k_2(m)} t + \phi_{k_2(m)}\right). \tag{4.31}$$

In Eqs. 4.28–4.31, A_{p_1}, A_{p_2}, A_{k_1} and A_{k_2} are the amplitudes while ϕ_{p_1}, ϕ_{p_2}, ϕ_{k_1} and ϕ_{k_2} are the phases at the various frequencies. ω_{k_1} and ω_{k_2} denote the angular frequencies of the uncorrelated harmonics k_1 and k_2, respectively. Note that $k_1(l) \neq k_2(m); \forall l, m$. N_l and N_m are the number of uncorrelated harmonics in $q_1(t)$ and $q_2(t)$, respectively. The uncorrelated harmonics should cover the entire bandwidth of interest.

The key idea in the VTFBI approach centres around the design of the correlated harmonic (Tan and Yap 2012). The frequency ω_0 of the correlated harmonic should be chosen where system gain is expected to be relatively large. The aim is to design this component such that it will equally excite all output directions, and by virtue of the relatively large gain, the effect of this component will be significant at the output. Limiting the analysis to the correlated harmonic alone for the moment,

$$Y_1(s) = G_{11}(s)P_1(s) + G_{12}(s)P_2(s), \tag{4.32}$$

$$Y_2(s) = G_{21}(s)P_1(s) + G_{22}(s)P_2(s). \tag{4.33}$$

Solving these simultaneously gives

$$P_1(s) = \frac{G_{22}(s)Y_1(s) - G_{12}(s)Y_2(s)}{G_{11}(s)G_{22}(s) - G_{12}(s)G_{21}(s)}, \tag{4.34}$$

$$P_2(s) = \frac{-G_{21}(s)Y_1(s) + G_{11}(s)Y_2(s)}{G_{11}(s)G_{22}(s) - G_{12}(s)G_{21}(s)}. \tag{4.35}$$

Equal excitation in all output directions is achieved if the relationship between $y_1(t)$ and $y_2(t)$ describes a circle in the output state-space. If there are output bounds, they can be taken into account during the design stage. Unequal output bounds for $y_1(t)$ and $y_2(t)$ can be easily accommodated via appropriate scaling.

Select

$$y_1(t) = \sin(\omega_0 t), \tag{4.36}$$

$$y_2(t) = \pm\cos(\omega_0 t), \tag{4.37}$$

where '±' denotes either '+' or '−'. The former and latter choices result in clockwise and counterclockwise rotation of the output state-space trajectory, respectively. This choice may be set arbitrarily in most cases. Taking the Laplace transform of Eqs. 4.36 and 4.37 leads to

$$Y_1(s) = \frac{\omega_0}{s^2 + \omega_0^2}, \tag{4.38}$$

$$Y_2(s) = \pm\frac{s}{s^2 + \omega_0^2}. \tag{4.39}$$

From Eqs. 4.34, 4.35, 4.38 and 4.39, the mathematical relationship between the two correlated inputs can be obtained and expressed as a ratio $H(s)$ in the form similar to that of a transfer function, where

$$\begin{aligned} H(s) = \frac{P_2(s)}{P_1(s)} &= \frac{-G_{21}(s)Y_1(s) + G_{11}(s)Y_2(s)}{G_{22}(s)Y_1(s) - G_{12}(s)Y_2(s)} \\ &= \frac{\pm sG_{11}(s) - \omega_0 G_{21}(s)}{\mp sG_{12}(s) + \omega_0 G_{22}(s)}. \end{aligned} \tag{4.40}$$

$H(s)$ is defined as the VTFBI (Tan and Yap 2012). The two key design parameters relate the magnitudes and phases between the two correlated inputs. These are given by

$$\frac{A_{p_2}}{A_{p_1}} = |H(s)||_{s=\mathrm{j}\omega_0} = \left|\frac{\pm\mathrm{j}G_{11}(\mathrm{j}\omega_0) - G_{21}(\mathrm{j}\omega_0)}{\mp\mathrm{j}G_{12}(\mathrm{j}\omega_0) + G_{22}(\mathrm{j}\omega_0)}\right|, \tag{4.41}$$

and

$$\phi_{p_2} - \phi_{p_1} = \angle H(s)|_{s=\mathrm{j}\omega_0} = \angle\left(\frac{\pm \mathrm{j}G_{11}(\mathrm{j}\omega_0) - G_{21}(\mathrm{j}\omega_0)}{\mp \mathrm{j}G_{12}(\mathrm{j}\omega_0) + G_{22}(\mathrm{j}\omega_0)}\right). \tag{4.42}$$

As the key design parameters depend on the estimates of the transfer functions $\hat{G}_{11}(\mathrm{j}\omega_0)$, $\hat{G}_{12}(\mathrm{j}\omega_0)$, $\hat{G}_{21}(\mathrm{j}\omega_0)$ and $\hat{G}_{22}(\mathrm{j}\omega_0)$, a priori tests in the form of sinusoidal perturbations at frequency ω_0 can be conducted for each input separately. The sensitivity of the VTFBI technique in terms of the amount of distortion to the ideal circular output state-space trajectory is analysed in Yap et al. (2017) with respect to any possible inaccuracy in the estimate of $H(s)$.

Now consider the design of the uncorrelated harmonics. The harmonics are selected such that there is no common excited harmonics between the inputs. According to the guideline given by Tan and Yap (2012), as far as is practical, set the amplitudes of the uncorrelated harmonics in $q_1(t)$ and $q_2(t)$ such that the output spectra due to these have slightly smaller magnitudes at frequencies close to ω_0, compared with the magnitudes of the output spectra caused by the correlated component. While the correlated harmonic serves to promote equal excitation in all output directions, the uncorrelated harmonics are utilised for direct estimation as output power appearing in these harmonics can be attributed to the effect of a single input, assuming linearity of the system. The uncorrelated harmonics are typically specified to have a uniform spectrum, for uniform excitation across the frequency range of interest. However, the spectrum can be shaped if necessary but this choice must be supported with more a priori information on the expected frequency response of the system.

The design using VTFBI has the following advantages when applied to multivariable ill-conditioned systems (Tan and Yap 2012):

1. The selection of a single correlated harmonic at a frequency where the system gain is high means that input power is more effectively utilised. This increases the estimation accuracy.
2. The method is insensitive to changes in the high and low gain directions with frequency, since the correlated harmonic is set at a single frequency. The design is not affected by the gain directions at other frequencies.
3. The technique is simple and can be easily understood by practising engineers.
4. The method is an open-loop method which does not require the implementation of feedback controllers.
5. The design is D-optimal (maximises the determinant of the information matrix) as shown by Theorem 4.1. This ensures that the maximum amount of information is collected from the experiment. Such a design is therefore efficient in terms of both time and cost. Maximising the determinant of the information matrix will in turn minimise the variance of the parameter estimates (Darby and Nikolaou 2014).

Theorem 4.1 *(D-optimality of VTFBI) (Yap et al. 2017)*
For a dynamic 2×2 subsystem selected at a frequency ω_0, the VTFBI approach achieves D-optimality subject to output variance constraints bounded based on

the maximum allowable output amplitudes $|y_i|_{max\_specified}$ *such that* $var(y_i) \leq \frac{(|y_i|_{max\_specified})^2}{2} = \frac{r^2}{2}$ *for* $i = 1, 2$.

Proof The covariance matrix of the output $\mathbf{C_y} = \mathbf{G}\mathbf{C_u}\mathbf{G}^T$ is given by $\mathbf{C_y} = \mathrm{E}[y_i y_j] - \mathrm{E}[y_i]\mathrm{E}[y_j]$, where $\mathbf{C_u}$ denotes the covariance matrix of the input and E denotes the expectation operator. Using Eqs. 4.36 and 4.37, but with the maximum and minimum amplitudes scaled to $\pm r$ to give $y_1(t) = r\sin(\omega_0 t)$ and $y_2(t) = \pm r\cos(\omega_0 t)$, leads to $\mathrm{E}[y_i] = \mathrm{E}[y_j] = 0$ and $\mathbf{C_y} = \mathrm{E}[y_i y_j] = \begin{bmatrix} r^2/2 & 0 \\ 0 & r^2/2 \end{bmatrix} = \mathrm{diag}(r^2/2, r^2/2)$. The diagonality at the maximum variance bounds of $\mathbf{C_y}$ ensures that $\det(\mathbf{C_y^{-1}})$ is minimised. Hence, the design is D-optimal. □

Example

Given that a 2×2 system has frequency responses $G_{11}(\mathrm{j}\omega_0) = 5\angle 0.3$ rad, $G_{12}(\mathrm{j}\omega_0) = 3\angle 2$ rad, $G_{21}(\mathrm{j}\omega_0) = 1\angle 1$ rad and $G_{22}(\mathrm{j}\omega_0) = 4\angle 2.6$ rad, find $p_1(t)$ and $p_2(t)$ to result in a clockwise circular trajectory in the output state-space. Set the amplitude of $p_1(t)$ to 1. Verify your answer using MATLAB simulation. Also, find the radius of the trajectory using MATLAB.

Solution

Converting the frequency responses at ω_0 from polar form to rectangular form gives $G_{11}(\mathrm{j}\omega_0) = 4.7767 + \mathrm{j}1.4776$, $G_{12}(\mathrm{j}\omega_0) = -1.2484 + \mathrm{j}2.7279$, $G_{21}(\mathrm{j}\omega_0) = 0.5403 + \mathrm{j}0.8415$ and $G_{22}(\mathrm{j}\omega_0) = -3.4276 + \mathrm{j}2.0620$. From Eqs. 4.41 and 4.42, $\frac{A_{p2}}{A_{p1}} = |H(s)||_{s=\mathrm{j}\omega_0} = \left|\frac{\mathrm{j}G_{11}(\mathrm{j}\omega_0) - G_{21}(\mathrm{j}\omega_0)}{-\mathrm{j}G_{12}(\mathrm{j}\omega_0) + G_{22}(\mathrm{j}\omega_0)}\right| = 1.307$ and $\phi_{p_2} - \phi_{p_1} = \angle H(s)|_{s=\mathrm{j}\omega_0} = \angle\left(\frac{\mathrm{j}G_{11}(\mathrm{j}\omega_0) - G_{21}(\mathrm{j}\omega_0)}{-\mathrm{j}G_{12}(\mathrm{j}\omega_0) + G_{22}(\mathrm{j}\omega_0)}\right) = 0.2655$ rad. From Eq. 4.28, let $A_{p_1} = 1$ and $\phi_{p_1} = 0$ so that $p_1(t) = \sin(\omega_0 t)$. Then $A_{p_2} = 1.307$ and $\phi_{p_2} = 0.2655$ so that $p_2(t) = 1.307\sin(\omega_0 t + 0.2655)$.

In MATLAB, the problem may be solved using the following codes:

```
%define the frequency responses and convert from polar
%form to rectangular form
G11=5*exp(j*0.3);
G12=3*exp(j*2);
G21=1*exp(j*1);
G22=4*exp(j*2.6);

%compute VTFBI
H=(j*G11-G21)/(-j*G12+G22);
Ap1=1;
phip1=0;
Ap2=Ap1*abs(H);
phip2=phip1+angle(H);


%arbitrarily set angular frequency to 1 rad for
%simulation and plotting
p1=Ap1*sin([0:0.01:2*pi]);
p2=Ap2*sin([0:0.01:2*pi]+0.2655);
```

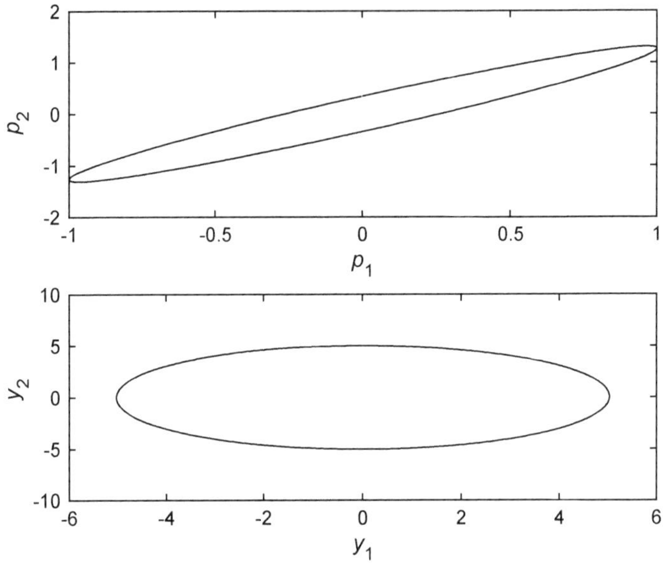

Fig. 4.12 Input and output state-space trajectories for VTFBI design using only the correlated component

```
%compute outputs
y11=abs(G11)*Ap1*sin([0:0.01:2*pi]+phip1+angle(G11));
y12=abs(G12)*Ap2*sin([0:0.01:2*pi]+phip2+angle(G12));
y21=abs(G21)*Ap1*sin([0:0.01:2*pi]+phip1+angle(G21));
y22=abs(G22)*Ap2*sin([0:0.01:2*pi]+phip2+angle(G22));
y1=y11+y12;
y2=y21+y22;
%find radius of circular trajectory
radius=max(y1)

%plot input and output state-space trajectories
subplot(2,1,1)
plot(p1,p2,'k');
xlabel('\itp\rm_{1}');ylabel('\itp\rm_{2}')
subplot(2,1,2)
plot(y1,y2,'k');
xlabel('\ity\rm_{1}');ylabel('\ity\rm_{2}')
```

The resulting plots are shown in Fig. 4.12. The radius of the output state-space trajectory is 5.03.

The extension to systems with higher dimension is considered next. For a system with υ outputs, decompose the system into ${}^{\upsilon}C_2$ smaller 2×2 subsystems across all combinations of $i, j = 1, 2, \ldots, \upsilon$ and $i \neq j$, such that

$$\begin{bmatrix} Y_i \\ Y_j \end{bmatrix} = \begin{bmatrix} G_{ik} & G_{il} \\ G_{jk} & G_{jl} \end{bmatrix} \begin{bmatrix} U_k \\ U_l \end{bmatrix} \tag{4.43}$$

with $k \neq l$. The VTFBI concept is applied to each subsystem. This requires that each subsystem be perturbed with a different frequency of the correlated component. For example, a 3×3 system can be decomposed into (Tan and Yap 2012)

$$\begin{bmatrix} Y_1(s) \\ Y_2(s) \end{bmatrix} = \begin{bmatrix} G_{11}(s) & G_{12}(s) \\ G_{21}(s) & G_{22}(s) \end{bmatrix} \begin{bmatrix} U_1(s) \\ U_2(s) \end{bmatrix}, \tag{4.44}$$

$$\begin{bmatrix} Y_1(s) \\ Y_3(s) \end{bmatrix} = \begin{bmatrix} G_{11}(s) & G_{13}(s) \\ G_{31}(s) & G_{33}(s) \end{bmatrix} \begin{bmatrix} U_1(s) \\ U_3(s) \end{bmatrix}, \tag{4.45}$$

$$\begin{bmatrix} Y_2(s) \\ Y_3(s) \end{bmatrix} = \begin{bmatrix} G_{22}(s) & G_{23}(s) \\ G_{32}(s) & G_{33}(s) \end{bmatrix} \begin{bmatrix} U_2(s) \\ U_3(s) \end{bmatrix}. \tag{4.46}$$

In this case, three different frequencies ω_0, ω_1 and ω_2 will be required for the correlated components. These can normally be set close to one another in a region where the process has a relatively high gain. From (4.44), find

$$u_1(t) = A_{p_1} \sin(\omega_0 t + \phi_{p_1}), \tag{4.47}$$

$$u_2(t) = A_{p_2} \sin(\omega_0 t + \phi_{p_2}) \tag{4.48}$$

such that (y_1, y_2) traces a circle when $u_3(t)=0$. From (4.45), find

$$u_1(t) = A_{p_3} \sin(\omega_1 t + \phi_{p_3}), \tag{4.49}$$

$$u_3(t) = A_{p_4} \sin(\omega_1 t + \phi_{p_4}) \tag{4.50}$$

so that (y_1, y_3) describes a circle when $u_2(t)=0$. Similarly, from (4.46), select

$$u_2(t) = A_{p_5} \sin(\omega_2 t + \phi_{p_5}), \tag{4.51}$$

$$u_3(t) = A_{p_6} \sin(\omega_2 t + \phi_{p_6}) \tag{4.52}$$

to make (y_2, y_3) follow a circular trajectory when $u_1(t)=0$. In Eqs. 4.47–4.52, A_{p_1} to A_{p_6} denote amplitudes while ϕ_{p_1} to ϕ_{p_6} represent phases. Finally, the correlated components in the signals are given by

$$u_1(t) = c_1 A_{p_1} \sin(\omega_0 t + \phi_{p_1}) + c_2 A_{p_3} \sin(\omega_1 t + \phi_{p_3}), \tag{4.53}$$

$$u_2(t) = c_1 A_{p_2} \sin(\omega_0 t + \phi_{p_2}) + c_3 A_{p_5} \sin(\omega_2 t + \phi_{p_5}), \tag{4.54}$$

$$u_3(t) = c_2 A_{p_4} \sin(\omega_1 t + \phi_{p_4}) + c_3 A_{p_6} \sin(\omega_2 t + \phi_{p_6}), \tag{4.55}$$

where c_1, c_2 and c_3 are scaling factors which can be chosen by the user.

Other choices of input pairs can be used instead of Eqs. 4.44–4.46; only the output pairs need to be selected exhaustively. The differences caused by the different possible choices are not significant; this eases the process of user selection (Yap et al. 2017).

Besides the extension described above which is based on system decomposition, another form of extension termed the direct extension is explained in Yap et al. (2017). The advantage of the direct extension is that only a single correlated component is needed but the method is limited to a maximum of six inputs and six outputs. A detailed case study on a simulated plasma etching reactor system is also presented in Yap et al. (2017), where various different designs are compared and evaluated.

4.3.2.1 Application Example

The system considered is based on a model of a real multizone tube furnace which can be used for applications in metallurgy and semiconductor fabrication (Yap and Tan 2011). Details of the results are described in Tan and Yap (2012). The furnace is constructed from a mullite tube of length 110 cm, an inner diameter of 65 mm and an outer diameter of 75 mm. It is divided horizontally into three zones of approximately equal length. Temperature control is achieved by adjusting the currents to the resistive coils. The coils are made from kanthal wires with a gauge of 17, wire resistance of approximately 1.397 Ω/m and an outer diameter of 1.15 mm.

In this example, the inputs $u_1(t)$ and $u_2(t)$ represent power in terms of the currents squared (measured in A^2) to the heating coils of the centre and right zones, respectively. The outputs $y_1(t)$ and $y_2(t)$ are the temperatures (measured in °C) at the left and centre zones, respectively. The dynamics of the system vary with the operating point. It is shown in Yap and Tan (2011) that the dynamics are slower when stepping the current up than when stepping the current down. This nonlinear distortion is mainly due to the effect of heat losses to the surrounding. For simplicity, a model linearised at $u_1 = u_2 = 10\text{A}^2$ is used here; this defines the reference zero for u_1 and u_2 to be at 10A^2. The inputs and outputs are related by

$$\begin{bmatrix} Y_1(s) \\ Y_2(s) \end{bmatrix} = \begin{bmatrix} \frac{3.70e^{-5s}}{146.30s+1} & \frac{0.86e^{-5s}}{(5.50s+1)(209.15s+1)} \\ \frac{29.15e^{-3s}}{39.50s+1} & \frac{5.27e^{-5s}}{130.46s+1} \end{bmatrix} \begin{bmatrix} U_1(s) \\ U_2(s) \end{bmatrix}, \tag{4.56}$$

where time is measured in minutes. The magnitudes of the FRFs are plotted in Fig. 4.13 where it can be seen that all four transfer functions exhibit lowpass characteristics. However, their magnitudes are significantly different.

Singular value decomposition applied to the steady-state transfer function matrix $\mathbf{G}(0)$ leads to

$$\mathbf{G}(0) = \begin{bmatrix} 0.1270 & 0.9919 \\ 0.9919 & -0.1270 \end{bmatrix} \begin{bmatrix} 29.8645 & 0 \\ 0 & 0.1865 \end{bmatrix} \begin{bmatrix} 0.9839 & -0.1787 \\ 0.1787 & 0.9839 \end{bmatrix}^T. \tag{4.57}$$

The singular values and the condition number are depicted in Fig. 4.14. The process is clearly ill-conditioned, with the condition number being larger than 100 across the entire frequency range of interest.

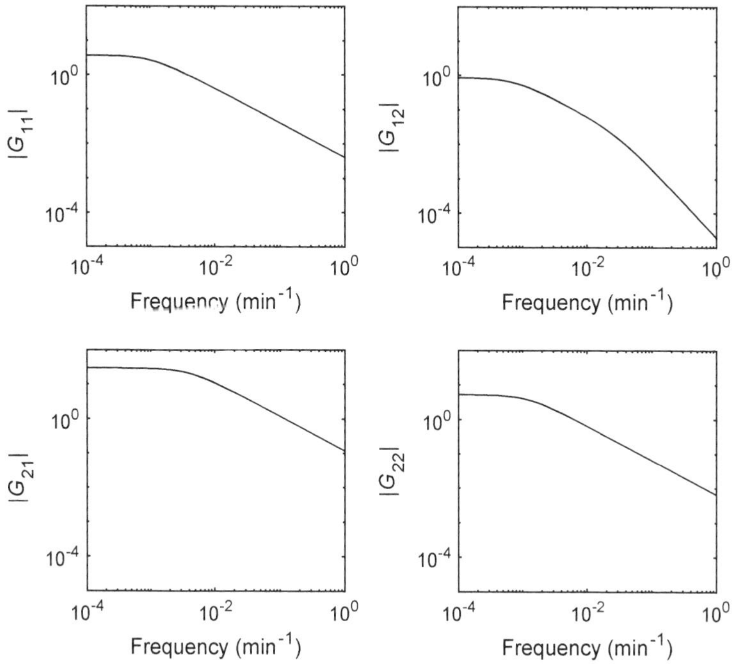

Fig. 4.13 Magnitudes of the FRFs of the multizone furnace

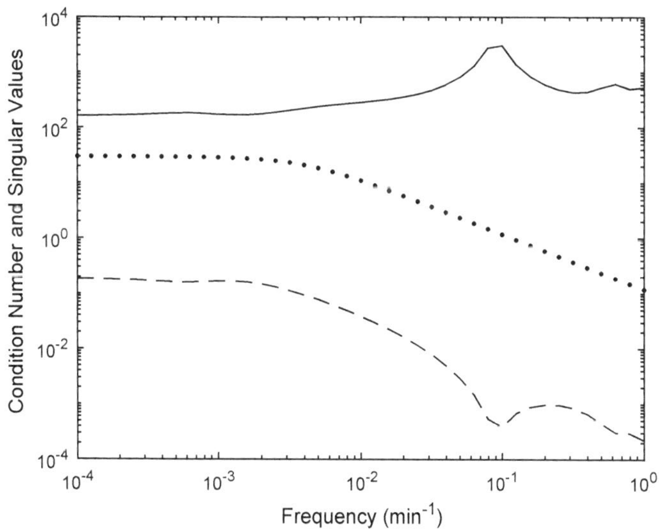

Fig. 4.14 Solid line: condition number; dotted line: $\sigma_1(s)$; dashed line: $\sigma_2(s)$ (Reproduced with permission from Tan and Yap © 2012 IET)

From Eq. 1.2, the sampling interval t_s should satisfy

$$t_s \leq \frac{\text{minimum time constant}}{5} = \frac{5.5\,\text{min}}{5} = 1.1\,\text{min} \tag{4.58}$$

whereas from Eq. 1.3, the measurement period T_N should meet

$$T_N \geq 5 \times \text{maximum time constant} = 5 \times 209.15\,\text{min} = 1045.75\,\text{min}. \tag{4.59}$$

In the simulations, these were set to $t_s = 0.2$ min and $T_N = 1500$ min, resulting in a sampling frequency of 5 min^{-1}, a frequency resolution of 6.67×10^{-4} min^{-1} and a signal period of 7500. A high sampling frequency was chosen so that the signals obtained approximate their analogue versions, as the transfer functions were to be identified in continuous-time.

For the VTFBI method, the uncorrelated components were selected at the following harmonics:

$$\text{In } u_1(t)\text{: } 3k + 1,\ 0 \leq k \leq 199, \quad k \in \mathrm{Z}; \tag{4.60}$$

$$\text{In } u_2(t)\text{: } 3k + 2,\ 0 \leq k \leq 199, \quad k \in \mathrm{Z}. \tag{4.61}$$

The highest harmonic was placed at 0.4 min^{-1} which is 8% of the sampling frequency. The choice was made considering a trade-off between using a larger number of harmonics for estimation and limiting the highest frequency used so that the effects of ZOH conversion can be considered negligible. The correlated harmonic was placed at a low frequency corresponding to harmonic number 3 ($\omega_0 = \frac{\pi}{250}$ rad min^{-1}) where the system gain is relatively large. Theoretically, from Eqs. 4.40 and 4.56, and arbitrarily selecting a counterclockwise rotation of the output state-space trajectory, the VTFBI is given by

$$\begin{aligned} H(s) &= \frac{P_2(s)}{P_1(s)} \\ &= -\frac{3.70s(39.50s+1)\mathrm{e}^{-5s} + 29.15\omega_0(146.30s+1)\mathrm{e}^{-3s}}{0.86s(130.46s+1)\mathrm{e}^{-5s} + 5.27\omega_0(5.50s+1)(209.15s+1)\mathrm{e}^{-5s}}. \\ &\quad \times \frac{(130.46s+1)(5.50s+1)(209.15s+1)}{(39.50s+1)(146.30s+1)} \end{aligned} \tag{4.62}$$

From Eq. 4.62, and substituting for the value of ω_0,

$$H(\mathrm{j}\omega_0) = 9.590\angle - 2.606. \tag{4.63}$$

At ω_0, $G_{11}(\mathrm{j}\omega_0) = 0.7456 - \mathrm{j}1.6030$, $G_{12}(\mathrm{j}\omega_0) = 0.0701 - \mathrm{j}0.2969$, $G_{21}(\mathrm{j}\omega_0) = 22.9335 - \mathrm{j}12.4822$ and $G_{22}(\mathrm{j}\omega_0) = 1.2792 - \mathrm{j}2.4280$. An a priori test using sequential perturbation of the inputs was carried out to estimate $\hat{H}(\mathrm{j}\omega_0)$ based on the estimates of $\hat{G}_{11}(\mathrm{j}\omega_0)$, $\hat{G}_{12}(\mathrm{j}\omega_0)$, $\hat{G}_{21}(\mathrm{j}\omega_0)$ and $\hat{G}_{22}(\mathrm{j}\omega_0)$. A sinusoidal signal of frequency ω_0 was applied to each input, one at a time. The signal had RMS amplitude of $5\mathrm{A}^2$

and a length of 3750 samples. The total length of the a priori test was 7500 samples, which corresponds to 1500 min.

For the modified zippered spectrum design, the uncorrelated harmonics were set at the same frequencies as those in the VTFBI approach, according to Eqs. 4.60 and 4.61. The correlated harmonics were placed at

$$\text{In } u_1(t) \text{ and } u_2(t)\text{: } 3k, \quad 1 \leq k \leq 200, \quad k \in \mathrm{Z} \tag{4.64}$$

The correlated components were designed based on an estimate of the low gain direction gathered from two step tests, applied to one input at a time. Each step test was of amplitude 5A^2, and comprised 3750 samples. These parameters were selected for fair comparison with the VTFBI technique in terms of the signal amplitude as well as the total duration of the a priori test. The ratio between the correlated harmonics and the uncorrelated harmonics was set such that the DFT magnitudes at the outputs due to the former and the latter were as nearly equal as possible, following the guideline in Rivera et al. (2009).

The process was subject to both input and output constraints. To satisfy input power constraints, both sets of signals were scaled such that

$$\mathrm{E}\left(\sqrt{u_1^2(t) + u_2^2(t)}\right) = 10\mathrm{A}^2 \tag{4.65}$$

The input signals are depicted in Figs. 4.15 and 4.16 for the VTFBI and modified zippered spectrum designs, respectively. While both signal sets have approximately the same amplitudes, the VTFBI design has a clearly observable correlated component between u_1 and u_2. The output constraints dictate that the deviations from the nominal values should be bounded between ±50 °C, in order to prevent over-exciting the nonlinearities as well as to promote plant-friendliness. This was found to be met in both designs.

The simulation was run 100 times. Zero-mean noise with RMS values of 0.01 and 0.02 °C was added to the outputs in two separate experiments. The singular values were first estimated employing the maximum likelihood estimator in the frequency domain, making use of data corresponding to the uncorrelated harmonics as the effects of the individual inputs can be easily decoupled at these points. The initial estimates were then improved via multidimensional unconstrained nonlinear minimisation in the time domain using data corresponding to both correlated and uncorrelated harmonics. The accuracy of the singular value estimates was measured based on its mean error defined by

$$\text{mean error in } \sigma_i(s) = \frac{\sum_{r=1}^{100} \left|\hat{\sigma}_{i,r}(s) - \sigma_i(s)\right|}{100}, \quad i = \{1, 2\}, \tag{4.66}$$

where $\hat{\sigma}_{i,r}(s)$ denotes the estimated value of $\sigma_i(s)$ at run r. The mean errors are plotted in Fig. 4.17.

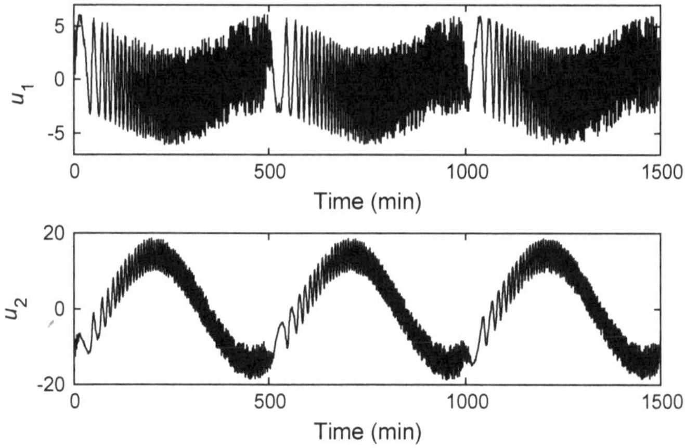

Fig. 4.15 Input signals using the VTFBI design

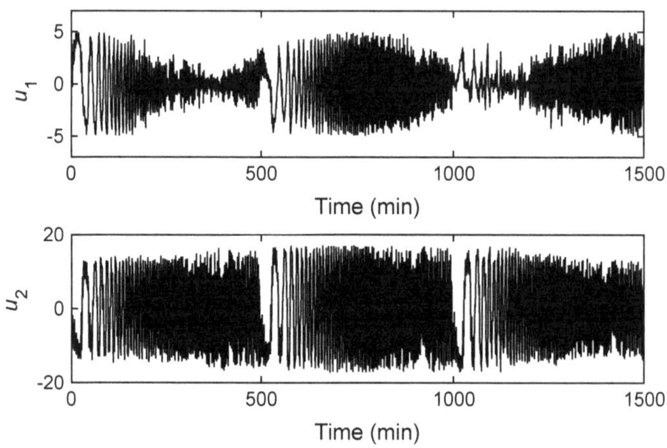

Fig. 4.16 Input signals using the modified zippered spectrum design

From Fig. 4.17, it can be seen that the VTFBI technique achieves more accurate estimates for both singular values and has greater robustness towards the effects of noise. This finding may be attributed to the following:

- The correlated harmonic was placed at a frequency where the gain is high. This increased the output SNR and improved the coverage of the output state-space.
- More power was distributed to the uncorrelated harmonics which were applied in the first stage estimation using maximum likelihood estimator. More accurate results in this stage mean better initial values for the second stage optimisation using nonlinear minimisation. In particular, the values of $\mathrm{E}\left(\sqrt{u_1^2(t)+u_2^2(t)}\right)$ and

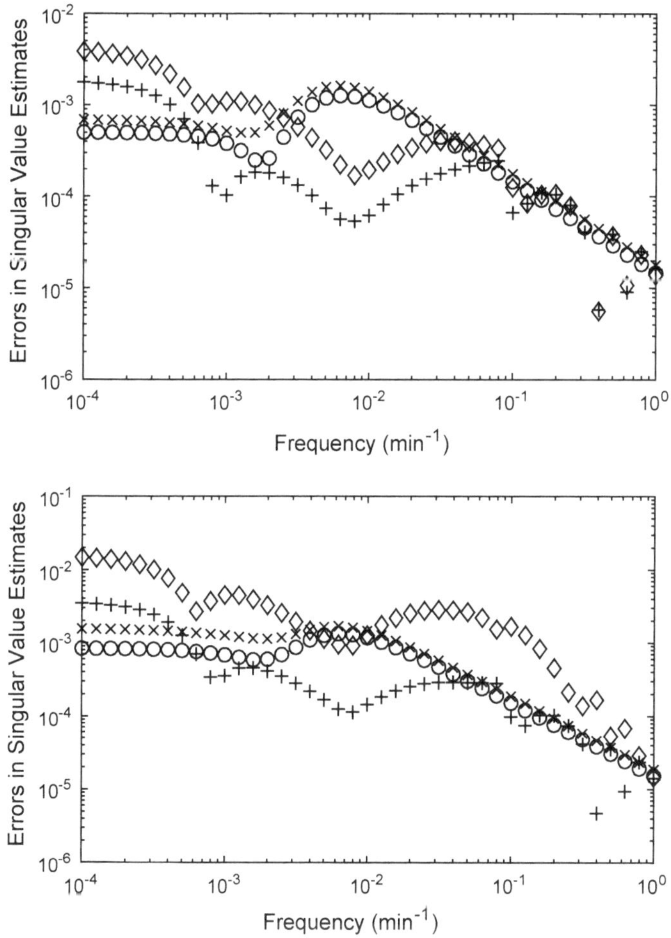

Fig. 4.17 Top: mean errors for noise RMS = 0.01 °C; bottom: mean errors for noise RMS = 0.02 °C. Circles: errors in $\sigma_1(s)$ using VTFBI; crosses: errors in $\sigma_1(s)$ using modified zippered spectrum; plusses: errors in $\sigma_2(s)$ using VTFBI; diamonds: errors in $\sigma_2(s)$ using modified zippered spectrum (Reproduced with permission from Tan and Yap © 2012 IET)

$\mathrm{E}\left(\sqrt{y_1^2(t) + y_2^2(t)}\right)$ for the VTFBI design, taking into account of only the uncorrelated components, were 1.9 times of those for the modified zippered spectrum design.

- Changes in the gain directions with respect to frequency have a negative impact on the modified zippered spectrum technique, but not on the VTFBI design.

According to Jacobsen et al. (1991), accurate estimates of the relative gain array (RGA) elements at steady-state, RGA(0), are important for closed-loop control. For the multizone furnace,

Table 4.4 Accuracy in the estimates of $\lambda_{11}(0)$

Technique	Mean absolute error (%) for noise RMS = 0.01 °C	Mean absolute error (%) for noise RMS = 0.02 °C	Standard deviation of error (%) for noise RMS = 0.01 °C	Standard deviation of error (%) for noise RMS = 0.02 °C
VTFBI	1.00	1.89	1.20	2.60
Modified zippered spectrum	2.06	7.03	3.29	9.57

Table 4.5 Accuracy in the estimates of the determinant of **G**(0)

Technique	Mean absolute error (%) for noise RMS = 0.01 °C	Mean absolute error (%) for noise RMS = 0.02 °C	Standard deviation of error (%) for noise RMS = 0.01 °C	Standard deviation of error (%) for noise RMS = 0.02 °C
VTFBI	0.98	1.93	1.20	2.75
Modified zippered spectrum	2.14	8.29	3.71	12.12

$$\mathrm{RGA}(0) = \mathbf{G}(0) \circ \left[\mathbf{G}^{-1}(0)\right]^T = \begin{bmatrix} -3.5 & 4.5 \\ 4.5 & -3.5 \end{bmatrix}, \tag{4.67}$$

where $\circ$ denotes the Schur product. The 1, 1 element of RGA(0), denoted by $\lambda_{11}(0)$, was identified using both methods. Results are summarised in Table 4.4 where it can be seen that the VTFBI method achieved more accurate estimation compared with the modified zippered spectrum approach. Similar results were obtained when the determinant of **G**(0) was identified, as shown in Table 4.5.

References

Barker HA, Tan AH, Godfrey KR (2014) Object-oriented creation of input signals for system identification. IET Control Theory Appl 8:821–829

Briggs PAN, Godfrey KR (1966) Pseudorandom signals for the dynamic analysis of multivariable systems. Proc Inst Electr Eng 113:1259–1267

Briggs PAN, Godfrey KR (1976) Design of uncorrelated signals. Electron Lett 12:555–556

Bruwer M-J, MacGregor JF (2006) Robust multi-variable identification: optimal experimental design with constraints. J Process Control 16:581–600

Chin CS, Munro N (2003) Control of the ALSTOM gasifier benchmark problem using H_2 methodology. J Process Control 13:759–768

Conner JS, Seborg DE (2004) An evaluation of MIMO input designs for process identification. Ind Eng Chem Res 43:3847–3854

Darby ML, Nikolaou M (2014) Identification test design for multivariable model-based control: an industrial perspective. Control Eng Pract 22:165–180

Jacobsen EW, Lundström P, Skogestad S (1991) Modelling and identification for robust control of ill-conditioned plants—a distillation case study. In: Proceedings of the American control conference, Boston, MA, 26–28 June, pp 242–248
Jin Q, Wang Z, Wang Q, Yang R (2013) Optimal input design for identifying parameters and orders of MIMO systems with initial values. Appl Math Comput 224:735–742
Jin Q, Wang Z, Yang R, Wang J (2014) An effective direct closed loop identification method for linear multivariable systems with colored noise. J Process Control 24:485–492
Kaigala GV, Jiang J, Backhouse CJ, Marquez HJ (2010) System design and modeling of a time-varying, nonlinear temperature controller for microfluidics. IEEE Trans Control Syst Technol 18:521–530
Kollár I (1994) Frequency domain system identification toolbox for use with MATLAB. The Math Works Inc., Natick
Li W, Lee JH (1996) Control relevant identification of ill-conditioned systems: estimation of gain directionality. Comput Chem Eng 20:1023–1042
Ljung L, Singh R (2012) Version 8 of the system identification toolbox. IFAC Proc Vol 45:1826–1831
MacWilliams J (1967) An example of two cyclically orthogonal sequences with maximum period. IEEE Trans Inf Theory 13:338–339
Martín CA, Rivera DE, Hekler EB (2016) An enhanced identification test monitoring procedure for MIMO systems relying on uncertainty estimates. In: Proceedings of the IEEE conference on decision and control, Las Vegas, NV, 12–14 December, pp 2091–2096
Misra S, Nikolaou M (2017) Adaptive design of experiments for model order estimation in subspace identification. Comput Chem Eng 100:119–138
Rasmussen KH, Jørgensen SB (1999) Parametric uncertainty modelling for robust control: a link to identification. Comput Chem Eng 23:987–1003
Rivera DE, Lee H, Mittelmann HD, Braun M (2009) Constrained multisine input signals for plant-friendly identification of chemical process systems. J Process Control 19:623–635
Roinila T, Huusari J, Vilkko M (2013) On frequency-response measurements of power-electronic systems applying MIMO identification techniques. IEEE Trans Indust Electron 60:5270–5276
Roinila T, Messo T, Santi E (2018) MIMO-identification techniques for rapid impedance-based stability assessment of three-phase systems in DQ domain. IEEE Trans Power Electron 33:4015–4022
Tan AH, Yap TTV (2012) Signal design for the identification of multivariable ill-conditioned systems using virtual transfer function between inputs. IET Control Theory Appl 6:394–402
Tan AH, Godfrey KR, Barker HA (2009) Design of ternary signals for MIMO identification in the presence of noise and nonlinear distortion. IEEE Trans Control Syst Technol 17:926–933
Tan AH, Barker HA, Godfrey KR (2015) Identification of multi-input systems using simultaneous perturbation by pseudorandom input signals. IET Control Theory Appl 9:2283–2292
Waller JB, Böling JM (2005) Multi-variable nonlinear MPC of an ill-conditioned distillation column. J Process Control 15:23–29
Yap TTV, Tan AH (2011) Identification of the static characteristics of a multizone furnace. In: Proceedings of the international conference on electrical, control and computer engineering, Kuantan, Malaysia, 21–22 June, pp 39–44
Yap TTV, Tan AH, Tan WN (2017) Identification of higher-dimensional ill-conditioned systems using extensions of virtual transfer function between inputs. J Process Control 56:58–68
Zhu Y, Stec P (2006) Simple control-relevant identification test methods for a class of ill-conditioned processes. J Process Control 16:1113–1120

Chapter 5
Signal Design for the Identification of Nonlinear and Time-Varying Systems

5.1 Objectives of Identification of Nonlinear Systems

For systems that have (or are suspected to have) significant nonlinearities, the perturbation signal must be tailored to the objective of the identification test. For example, if the objective is to identify the underlying linear dynamics of a system, the effects of nonlinearities can be minimised by suppressing the integer multiples of certain harmonics in the perturbation signal and then removing the power at the output corresponding to the suppressed harmonics. Alternatively, it is possible to take into account the effects of nonlinearities and yet model the system using a linear model if the best linear approximation of the system is estimated. If instead, the aim is to identify the nonlinearities themselves, dedicated signals and methods are available. Some general guidelines are summarised in Table 5.1.

Further to the design of the individual period of the signal, experiment design in terms of the number of periods applied as well as the number of different phase realisations of the signal plays an important role in determining the quality of the final estimated model.

5.2 Identification of the Best Linear Approximation

For perturbation using normalised random signals, the effects of nonlinear distortions on the FRF measurements can be explained by splitting the FRF $G(\mathrm{j}\omega)$, at an angular frequency ω into three components

$$G(\mathrm{j}\omega) = G_{\mathrm{RLD}}(\mathrm{j}\omega) + G_{\mathrm{S}}(\mathrm{j}\omega) + N_{\mathrm{G}}(\mathrm{j}\omega), \tag{5.1}$$

where $G_{\mathrm{RLD}}(\mathrm{j}\omega)$ is the related linear dynamics, $G_{\mathrm{S}}(\mathrm{j}\omega)$ represents the stochastic nonlinear contributions and $N_{\mathrm{G}}(\mathrm{j}\omega)$ is the error due to output noise (Schoukens

A. H. Tan and K. R. Godfrey, *Industrial Process Identification*, Advances in Industrial Control, https://doi.org/10.1007/978-3-030-03661-4_5

Table 5.1 Favourable signal properties for different identification objectives

Objectives	Favourable signal properties
To minimise effects of nonlinearities or to allow their detection	Signals with even harmonics suppressed which allow the elimination of the effects of even-order nonlinearities as well as their detection at the suppressed harmonics in the output Signals with harmonic multiples of three suppressed which allow the reduction of the effects of odd-order nonlinearities as well as their detection at the suppressed harmonics in the output
To obtain best linear approximation of nonlinear system	Signals that resemble those that the system will encounter in practice in terms of the amplitude distribution. If this is not readily known, signals with Gaussian amplitude distribution should be used
To estimate the nonlinearities by first estimating the best linear approximation	Signals with Gaussian amplitude distribution
To estimate specific forms of nonlinearities	Depends on the specific type of nonlinearity. For example, for the identification of Volterra kernels, no interharmonic distortion (NID) multisines may be applied (Evans et al. 1996; Tan and Godfrey 2002). For the identification of Hammerstein models with a linear pathway in parallel with a pathway consisting of a quadratic nonlinearity in series with a second linear block, truncated PRML signals may be utilised (Tan 2007)

et al. 2001). For the class of normally distributed signals, $G_{\mathrm{RLD}}(\mathrm{j}\omega)$ can be further decomposed into two components

$$G_{\mathrm{RLD}}(\mathrm{j}\omega) = G_0(\mathrm{j}\omega) + G_{\mathrm{B}}(\mathrm{j}\omega), \tag{5.2}$$

where $G_0(\mathrm{j}\omega)$ is the underlying linear dynamics and $G_{\mathrm{B}}(\mathrm{j}\omega)$ is the bias caused by nonlinear distortions. $G_{\mathrm{RLD}}(\mathrm{j}\omega)$ can be considered the best linear approximation ($=G_{\mathrm{BLA}}(\mathrm{j}\omega)$) of the nonlinear system. A linear filter with a transfer function of $G_{\mathrm{BLA}}(\mathrm{j}\omega)$ will minimise the squared differences between the output of the filter and the output of the actual nonlinear system. It gives the best possible fit of a linear model to a nonlinear system in the least squares sense.

The concept of best linear approximation is very useful when a nonlinear process is to be linearised around its operating point. From Eq. 5.1, it allows the nonlinear system to be modelled as a linear system with the stochastic nonlinear contributions treated as an output noise source. As such, linear control theory can be applied to the linearised model, which greatly simplifies analysis. However, the identification needs to be done with care as the best linear approximation is dependent on the perturbation signal applied. To see this, we consider the estimation of a static nonlinearity $y = u^3$. Four signals, all with equal signal power, are used to excite the system, which are given as follows:

- Signal A—a binary signal with levels symmetrical around zero and having zero mean,
- Signal B—a ternary signal with levels symmetrical around zero and having zero mean, plus having an equal probability of being at any of the three levels,
- Signal C—a signal with Gaussian distribution, and
- Signal D—a signal with uniform distribution.

From the results shown in Fig. 5.1, the best static linear approximation depends highly on the signal used. The reason is that different signals do not equally excite the different parts of the nonlinearity.

It is therefore highly recommended to apply perturbation signals that are as similar as possible in terms of the amplitude distribution and frequency content to the signals which the system will likely encounter in practice. A good choice would be several realisations of random-phase multisines with the same power spectrum but different phase in each realisation. The different realisations will help average out the effects of the stochastic nonlinear contributions. When M realisations are utilised, the standard deviation of the FRF due to the nonlinear distortions and the disturbing noise will be reduced by $\sqrt{M}$.

For the identification of nonlinear systems of the Wiener-Hammerstein structure of Fig. 5.2 where $G(z)$ and $H(z)$ represent linear dynamics, perturbation signals with Gaussian amplitude distribution have the advantage that the G_{BLA} will be a scaled

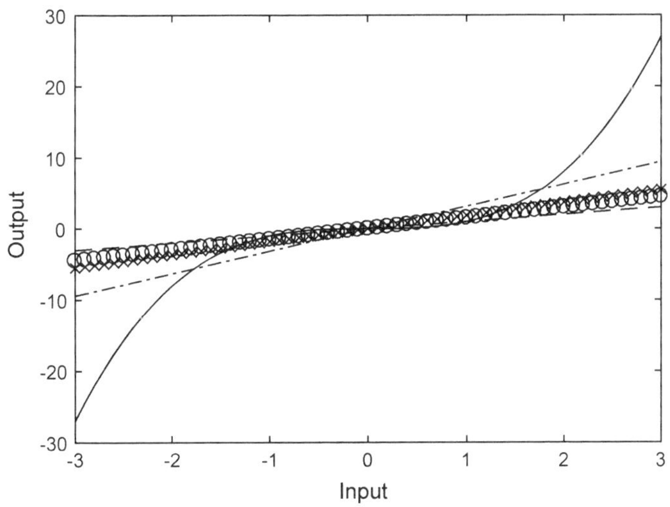

Fig. 5.1 Static nonlinearity $y = u^3$ (solid line) and the best linear approximation using Signal A (dashed line), Signal B (circles), Signal C (dashed-dotted line) and Signal D (crosses)

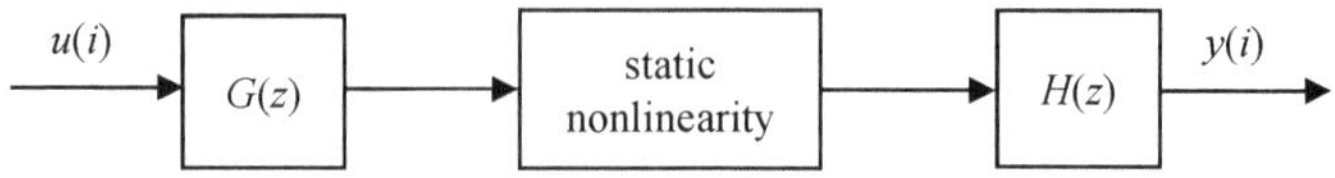

Fig. 5.2 Wiener-Hammerstein system structure

version of the product of the linear subsystems. Random-phase multisines can be applied to advantage in this situation as they resemble Gaussian noise asymptotically for sufficiently large N, while they also have the benefits of a deterministic signal (Schoukens et al. 2016). Using a random-phase multisine,

$$G_{\mathrm{BLA}}(z) = cG(z)H(z) + O\left(F^{-1}\right), \tag{5.3}$$

where c is a constant that depends on the higher order Volterra kernels and the power spectrum of the signal. There is an additional bias term which depends on the number of excited frequencies in the multisines used in the identification represented by F (Pintelon and Schoukens 2012). If the system is Wiener ($H = 1$) or Hammerstein ($G = 1$), it is straightforward to estimate the static nonlinearity once G_{BLA} is known. In the general Wiener-Hammerstein case, the situation is more involved and there is rich literature on this for the interested reader; see, for example, Schoukens et al. (2014), Ase and Katayama (2015), and Vanbeylen (2015). There is also a Special Section in Control Engineering Practice 2012 (volume 20, issue 11) on Wiener-Hammerstein System Identification Benchmark.

When non-Gaussian inputs are used, the best linear approximation obtained has a bias compared with the Gaussian case. The amount of bias depends on both the form of the nonlinearities and the higher order moments of the input sequence. To minimise the bias, moment matching can be performed, as proposed by Wong et al. (2013). The kth moment of a periodic random sequence $u(i)$ is defined by

$$\mathrm{M}_k \triangleq \mathrm{E}[u^k], \tag{5.4}$$

where E denotes the expectation operator. For discrete multilevel sequences with levels symmetrical around zero and having zero mean, $\mathrm{M}_k = 0$ for odd k. The moment matching attempts to match the even moments. For signals with m levels, the moment of degrees 2, 4, 6, … , $2(m - 1)$ can be matched. It should, however, be noted that if the system is Hammerstein with the highest order of the nonlinearity given by r, the input signal should have at least $r + 1$ different levels for it to be persistently exciting.

Example

Determine the amplitude distribution of a symmetrical ternary signal with moments matched as far as possible to a Gaussian input with zero mean and variance of 2. Compare the G_{BLA} obtained using the ternary signal and the Gaussian signal on a Chebyshev filter followed by a static nonlinearity of the form $f(x) = x + 0.5x^3$.

Table 5.2 Moments M_2 and M_4 for Gaussian and ternary inputs

Moments	Gaussian input	Ternary input
M_2	$\sigma^2 = 2$	$a^2 p$
M_4	$3\sigma^4 = 12$	$a^4 p$

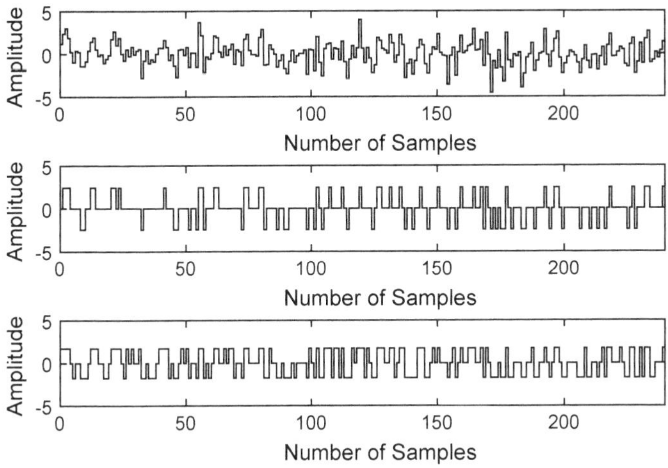

Fig. 5.3 Top: Gaussian input; middle: ternary signal matched to M_2 and M_4; bottom: ternary signal matched to M_2 only

Additionally, compare the results also with the G_{BLA} obtained using a ternary signal with uniform distribution (equal probability of being at any of the three levels).

Solution

For a ternary signal, moments of degrees 2 and 4 can be matched. Let the ternary signal have levels $\pm a$ with probability $p/2$ and level 0 with probability $1 - p$. The moments are given in Table 5.2.

Matching these give $a = \sqrt{6}$ and $p = 1/3$. For a ternary signal with uniform distribution, M_2 can still be matched but not M_4, giving $a = \sqrt{3}$ and $p = 2/3$.

In the simulation, the Gaussian input was generated using a random-phase multisine with $N = 240$. The ternary signals were generated from the Gaussian input by setting the lowest p bits in the signal to $-a$, the highest p bits to $+a$ and the rest to 0. This can be easily done with the help of the `sort` function in MATLAB®. (MATLAB® is a registered product of The MathWorks, Inc.) The signals are plotted in Fig. 5.3.

Results of the best linear approximation are shown in Fig. 5.4. The effect of averaging 100 realisations of each type of signal can be clearly observed, where the effects of the stochastic nonlinear contributions have been averaged out. The G_{BLA} obtained for the ternary signal which matched M_2 and M_4 is very close to that

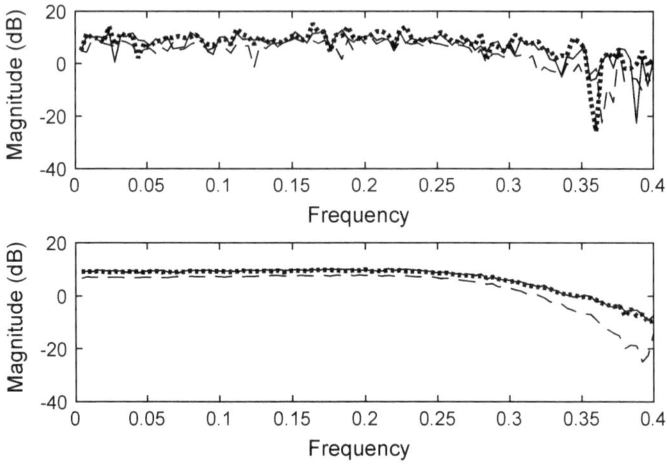

Fig. 5.4 G_{BLA} measured using Gaussian input (solid line), ternary signal matched to M_2 and M_4 (dotted line), and ternary signal matched to M_2 only (dashed line). Top: 1 realisation; bottom: average of 100 realisations

obtained using the Gaussian signal, but the bias is larger for the ternary signal which matched only M_2.

5.3 Identification of Volterra Kernels

Volterra functional representation is a generalisation of the power series to nonlinear systems with memory. For a causal, stable and time-invariant system, the time domain output $y(t)$ of a Volterra nonlinear system is given by

$$y(t) = \int_0^t h_1(\tau_1)u(t-\tau_1)\mathrm{d}\tau_1 + \int_0^t\int_0^t h_2(\tau_1,\tau_2)u(t-\tau_1)u(t-\tau_2)\mathrm{d}\tau_1\mathrm{d}\tau_2 + \cdots$$
$$+ \int_0^t\int_0^t \cdots \int_0^t h_n(\tau_1,\tau_2,\ldots,\tau_n)u(t-\tau_1)u(t-\tau_2)\ldots u(t-\tau_n)\mathrm{d}\tau_1\mathrm{d}\tau_2\ldots\mathrm{d}\tau_n, \tag{5.5}$$

where $h_1(\tau_1)$, $h_2(\tau_1, \tau_2)$ and $h_n(\tau_1, \tau_2, \ldots, \tau_n)$ are linear kernel, second-order kernel and nth-order kernel, respectively; u is the input; and τ is a time variable with the subscript denoting the dimension. For most practical systems, the effect of the higher order kernels decreases with increasing n, thus allowing a truncated series to be used (Weiss et al. 1998). An important advantage of the Volterra kernel is that it is valid in both time and frequency domains. However, identification in the frequency domain is more convenient since convolutions in the time domain become multiplications in the frequency domain. The n-dimensional kernel in the frequency domain is defined through

$$Y_n(s_1, s_2, \ldots, s_n) = H_n(s_1, s_2, \ldots, s_n)U(s_1)U(s_2)\ldots U(s_n), \quad (5.6)$$

where U, Y and H are the Laplace transforms of u, y and h, respectively; and s is the Laplace operator with the subscripts representing the frequencies of the particular dimensions. The Volterra kernel possesses symmetry (Zhang and Billings 2017) which is given by

$$H_n(s_1, s_2, \ldots, s_n)_{\text{sym}} = \frac{1}{n!}\left\{H_n(s_1, s_2, \ldots, s_n)_{\text{asym}} + \cdots + H_n(s_n, s_{n-1}, \ldots, s_1)_{\text{asym}}\right\}, \quad (5.7)$$

where a symmetrical kernel is obtained by making all possible combinations of the arguments of the asymmetric kernel. For example, the second-order Volterra kernel is symmetrical along the $f_1 = f_2$ and $f_1 = -f_2$ diagonals. Here, f_1 and f_2 are the values of the first and second dimensions of the input harmonics, respectively. Only a symmetrical kernel is guaranteed to be unique for a given system (Weiss et al. 1996).

The Volterra kernels can be readily measured using NID multisines (Evans et al. 1996). To understand the concept, assume that a signal with harmonics 1 and 3 is used to perturb a system with quadratic nonlinearity. Noting that negative frequencies need to be taken into account, the power at the output will appear at the harmonics as shown in Table 5.3.

Based on the symmetry in Eq. 5.7, the power at harmonic 6 will be able to yield the kernel $H_2(f_1 = 3, f_2 = 3)_{\text{sym}}$ and the power at harmonic 4 will be able to yield $H_2(f_1 = 1, f_2 = 3)_{\text{sym}}$, after applying Eq. 5.6. However, the power at harmonic 2 cannot be utilised in the estimation as it is contributed by $H_2(f_1 = 1, f_2 = 1)_{\text{sym}}$ and $H_2(f_1 = -1, f_2 = 3)_{\text{sym}}$. The effects of the two cannot be separated. In other words, interharmonic distortion has occurred. The power at harmonic 0 can never be utilised, as it is theoretically not possible to avoid interharmonic distortion here.

The design of NID multisines for second-order Volterra kernels aims to eliminate all interharmonic distortion except at harmonic 0. This can be done using the search procedure proposed by Evans et al. (1996), which seeks to maximise the measurement points while maintaining a near-even harmonic spacing and minimising the highest harmonic. A similar procedure was proposed by Han et al. (2014). It is interesting to note that if all the harmonics are chosen to be odd, the contribution of the linear term will not fall on the measurements points of the second-order kernel. An example of such a set of 10 harmonics, designed by Evans et al. (1996), is [3 13 25 43 57 77 119 155 203 227]. The measurement points are shown in Fig. 5.5. This signal achieves the theoretical maximum number of measurement points given by

$$\text{Maximum points for second-order kernel} = v^2 - v, \quad (5.8)$$

Table 5.3 Power at the output for second-order nonlinearity measured with a signal with harmonics 1 and 3

Harmonics at input f_1	Harmonics at input f_2	Harmonics at output
−3	−3	−6
−3	−1	−4
−3	1	−2
−3	3	0
−1	−3	−4
−1	−1	−2
−1	1	0
−1	3	2
1	−3	−2
1	−1	0
1	1	2
1	3	4
3	−3	0
3	−1	2
3	1	4
3	3	6

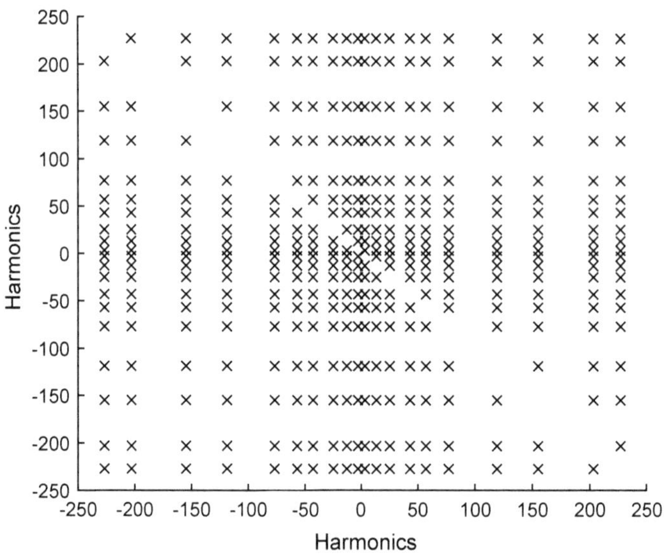

Fig. 5.5 Second order kernel coverage plot for signal with harmonic set [3 13 25 43 57 77 119 155 203 227], showing the positions in the frequency plane where the kernel can be estimated (© [2002] IEEE. Reprinted, with permission, from Tan and Godfrey (2002))

where $v = 2 \times$ number of harmonics in the signal. Equation 5.8 is derived by considering that f_1 and f_2 can each take v values, resulting in v^2 points. However, there are v points of the form $(f_1, f_2 = -f_1)$ which cannot be measured as their output power appears at harmonic 0.

For third-order Volterra kernels, the theoretical maximum number of measurement points is given by

$$\text{Maximum points for third-order kernel} = v^3 - 3v^2 + 3v. \tag{5.9}$$

Equation 5.9 is derived by considering that f_1, f_2 and f_3 can each take v values, resulting in v^3 points. However, for every value of f_1, the following combinations of harmonics $(f_1, f_2 = f_1, f_3 = -f_1)$, $(f_1, f_2 = -f_1, f_3 = f_1)$ and $(f_1, f_2 = -f_1, f_3 = -f_1)$ will result in output power at harmonic f_1 or $-f_1$, overlapping with the linear contribution and hence cannot be measured. There are a total of $3v$ such points. Further to this, there are $v(v - 2)$ combinations with $(f_1, f_2 = -f_1, f_3 \neq \pm f_1)$, $v(v - 2)$ combinations with $(f_1, f_2 \neq \pm f_1, f_3 = -f_1)$ and $v(v - 2)$ combinations with $(f_1, f_2 \neq \pm f_1, f_3 = -f_2)$ which cannot be measured as they will also generate power at the output coinciding with the linear contribution. An example of a harmonic set which achieves the theoretical maximum number of measurement points is given by Evans et al. (1996) as [241 451 663 877 1095 1319 1581 1817 2109 2347].

In general, the higher the order of the kernel, the sparser the harmonics are in the signal.

In the recent work by Han et al. (2014), signals which do not have interharmonic distortion for both second- and third-order kernels were designed. An example is an 8-harmonic signal with harmonic set given by [3 8 17 41 78 195 322 588]. The measurement points of second-order Volterra kernel are shown in Fig. 5.6. (Those of the third-order kernel are not shown here but they can be plotted using the MATLAB codes given later in this section.) It can be seen that the distribution of points is much less uniform compared with that in Fig. 5.5 due to the additional requirement of having no interharmonic distortion for third-order kernels.

MATLAB codes for evaluating the performance of sets of harmonics for the estimation of second- and third-order Volterra kernels are given below. These create a MATLAB function `perfmeasure`:

```
function [numpts,ratio_numpts,IHD,valid]
= perfmeasure(freq,order)

%This function evaluates the performance of a harmonic
%vector in terms of its usefulness for Volterra kernel
%measurements.
%freq - frequency vector
%order - order of nonlinearity (2 or 3)
%numpts - number of measurement points
%ratio_numpts - ratio between the number of measurement
%points to the theoretical maximum
%IHD - matrix of points
%The value of each element of the matrix shows the
%combined frequency at that point.
```

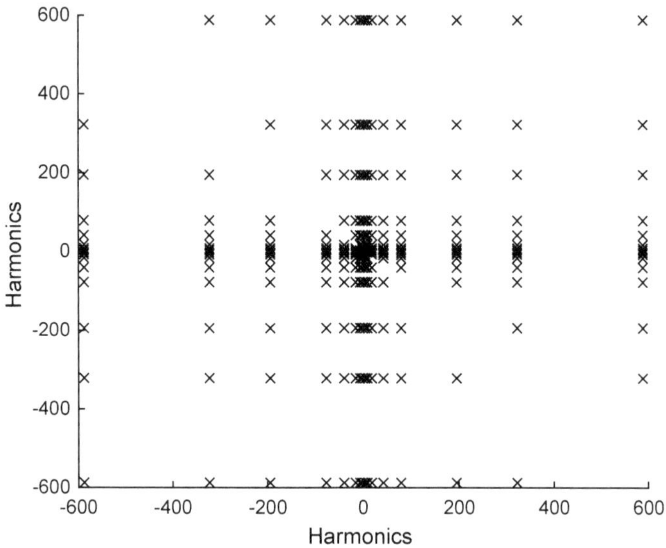

Fig. 5.6 Second order kernel coverage plot for signal with harmonic set [3 8 17 41 78 195 322 588], showing the positions in the frequency plane where the kernel can be estimated

```
%The matrix indices correspond to the harmonic elements
%giving the combined frequency value.
%valid - matrix showing the validity of each point in
%IHD (whether there is interharmonic distortion) for
%the measurement of Volterra kernel

%order of nonlinearity is 2
if order==2

%form matrix of frequency points
veclength=2*length(freq);
f=[-freq freq];f=sort(f);f1=f;f2=f;
for i=1:veclength
   for j=1:veclength
      IHD(i,j)=f1(i)+f2(j);
   end
end

%find points with interharmonic distortion
valid=ones(veclength,veclength);
for i=1:veclength
   for j=i:veclength
      a=IHD(i,j);
      [indexi,indexj]=find(IHD==a);
       if length(indexi)>2
          for b=1:length(indexi)
            valid(indexi(b),indexj(b))=0;
          end
       end
```

```
      end
   end

   %plot 2-dimensional plot
   figure;hold
   for i=1:veclength
      for j=1:veclength
          if valid(i,j)==1
             plot(f1(i),f2(j),'kx')
          end
      end
   end
   xlabel('f1');ylabel('f2')

   %find number of measurement points
   [c,d]=find(valid==1);
   numpts=length(c);
   %theoretical maximum value of numpts is
   %veclength^2-veclength
   %exclude points that will fall on harmonic zero
   ratio_numpts=numpts/(veclength^2-veclength);

   else
   %order of nonlinearity is 3

   %form matrix of frequency points
   veclength=2*length(freq);
   f=[-freq freq];f=sort(f);f1=f;f2=f;f3=f;
   for i=1:veclength
      for j=1:veclength
         for k=1:veclength
            IHD(i,j,k)=f1(i)+f2(j)+f3(k);
         end
      end
   end

   %find points with interharmonic distortion
   valid=ones(veclength,veclength,veclength);
   for i=1:veclength
      for j=i:veclength
            for k=j:veclength
               a=IHD(i,j,k);
               [indexi,indexj,indexk]=find(IHD==a);
                if length(indexi)>6 | (length(indexi)>3
   & (i==j|i==k|j==k))
                   for b=1:length(indexi)
                      for r=1:length(IHD)
                         [p,q]=find(IHD(:,:,r)==a);
                         for s=1:length(p)
                            valid(p(s),q(s),r)=0;
                         end
                      end
                   end
               end
            end
      end
```

```
end

%plot 3-dimensional plot
figure;hold
for i=1:veclength
   for j=1:veclength
       for k=1:veclength
          if valid(i,j,k)==1
              plot3(f1(i),f2(j),f3(k),'kx')
          end
       end
   end
end
xlabel('f1');ylabel('f2'),zlabel('f3')

%find number of measurement points
[c,d,e]=find(valid==1);
numpts=length(c);
%theoretical maximum value of numpts is
%veclength^3-3*veclength^2+3*veclength
%exclude points that will fall on the original test
%frequencies that will overlap with linear contribution
ratio_numpts=numpts/(veclength^3-3*veclength^2+3*veclength);
end
```

The function can be called from the MATLAB command window. For example, to check the performance of a set of three harmonics [3 5 7] for the measurement of the second-order Volterra kernel, type `[numpts,ratio_numpts,IHD,valid]=perfmeasure([3 5 7],2)` in the MATLAB command window. This will give

```
numpts = 16

ratio_numpts = 0.5333

IHD =
   -14   -12   -10    -4    -2     0
   -12   -10    -8    -2     0     2
   -10    -8    -6     0     2     4
    -4    -2     0     6     8    10
    -2     0     2     8    10    12
     0     2     4    10    12    14

valid =
     1     1     0     1     0     0
     1     0     1     0     0     0
     0     1     1     0     0     1
     1     0     0     1     1     0
     0     0     0     1     0     1
     0     0     1     0     1     1
```

where the kth row of `IHD` corresponds to f_1 being equal to the kth element in [−7 −5 −3 3 5 7] and the lth column of `IHD` corresponds to f_2 being equal to the lth element in [−7 −5 −3 3 5 7].

The output at harmonic 2 cannot be used to measure the second-order Volterra kernel because the contributions from $H_2(f_1 = -5, f_2 = 7)_{sym}$ and $H_2(f_1 = -3, f_2 = 5)_{sym}$ cannot be separated. The corresponding values of `valid` in (row, column) = (2, 6), (3, 5), (5, 3) and (6, 2) are equal to zero, showing that these points are not valid. The output at harmonic 10 cannot be used to measure the second-order Volterra kernel because the contributions from $H_2(f_1 = 3, f_2 = 7)_{sym}$ and $H_2(f_1 = 5, f_2 = 5)_{sym}$ cannot be separated. Similarly, the corresponding values of `valid` in (row, column) = (4, 6), (5, 5) and (6, 4) are equal to zero. The number of valid measurement points is 16, which is the number of ones in `valid`. The theoretical maximum according to Eq. 5.8 is $6^2 - 6 = 30$. This gives a ratio of 16/30 = 0.5333.

5.4 Quantification of Effects of Nonlinearities, Noise and Time Variation

5.4.1 Method Based on Frequency Domain Analysis

A carefully designed perturbation signal can aid the quantification of nonlinearities, noise and time variation. In particular, the application of several periods of a periodic signal with harmonic suppression facilitates detailed analysis of the system which is particularly convenient in the frequency domain. Comparatively little research reported in the literature capitalises on this, exceptions being the papers by Lataire et al. (2012) and Pintelon et al. (2013) which provide detailed frequency domain analysis of the output spectrum for single-input systems. A recent work by Tan (2018) describes results for a dual-stage hard disk drive with two inputs and a single output.

Consider a simple case where a single sinusoid with frequency ω perturbs a system with parallel linear and quadratic paths, as shown in Fig. 5.7, where A represents a gain term and ϕ represents a phase shift. At the output of the linear block, the frequency is the same as that of the input. However, at the output of the quadratic nonlinearity, the power appears at frequencies 0 and 2ω. This can be generalised such that for input signals with power only at the odd harmonics, the contributions of even-order nonlinearities will appear only at the even harmonics at the output. For an input with harmonics at integer multiples of three suppressed as well, the contributions of odd-order nonlinearities appear at the odd harmonics and will be easily detected based on the power at the odd multiples of three. This is illustrated in Fig. 5.8, where a linear system with odd- and even-order nonlinearities is perturbed by a signal with harmonic multiples of two and three suppressed, the excited harmonics being at 1, 5, 7, 11, ….

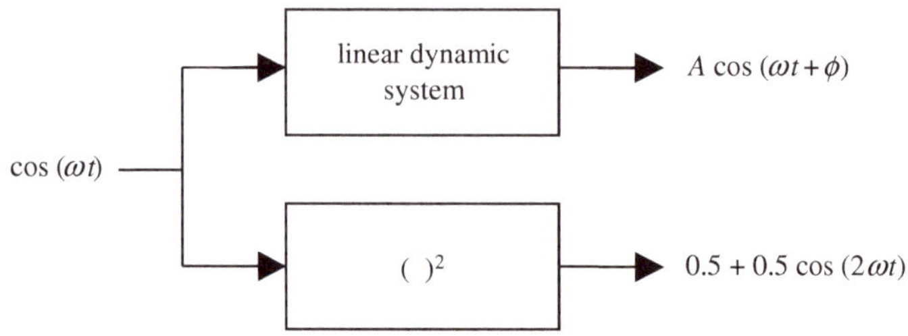

Fig. 5.7 Parallel system perturbed with a single sinusoid

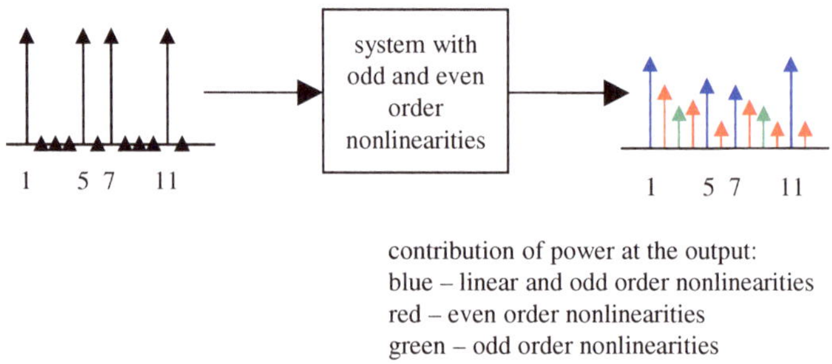

Fig. 5.8 Detection of even and odd-order nonlinearities using input signal with harmonic multiples of two and three suppressed

A step-by-step methodology which is applicable also to multi-input systems is as follows (Cham et al. 2017):

Step 1: Perturb all the inputs u_j ($j = 1, 2, \ldots, r$) of the system simultaneously using a set of signals of period N which are uncorrelated with one another. (For the design of uncorrelated signal sets, see Sect. 4.1.) Suppress the even harmonics in all the signals so that the effects of even-order nonlinearities can be filtered out at the system output (Pintelon and Schoukens 2012). This ensures that they will not corrupt the linear estimates. Harmonic multiples of three can also be suppressed if longer experimentation time can be tolerated. This is helpful for reducing the effects of odd-order nonlinearities on the linear estimates. Measure P periods ($P \geq 2$) of the output(s) y_i ($i = 1, 2, \ldots, s$) after the transient has sufficiently decayed.

Step 2: Compute the N-point DFTs U_j and Y_i corresponding to u_j and y_i, respectively. Plot the measured FRFs defined by $G_{ij}(z^{-1}) = Y_i(z^{-1})/U_j(z^{-1})$ for all the individual steady-state periods ($1, 2, \ldots, P$) and the averaged period. Visually check for differences between the plots. Such differences may point to the presence of time-varying components.

Step 3: Plot the ($N \times P$)-point DFT of the output using the whole measurement data of $N \times P$ points. Since both linear and nonlinear terms are periodic with the same period N as the inputs, the power at the output due to these will appear at harmonics which are integer multiples of P. Harmonic suppression thus provides an indication of the contribution of the nonlinear terms through the power appearing at the lines

$P \times$ (non-excited harmonics). In contrast, the contribution from disturbances will appear at all the harmonics, and its relative significance can be judged from power at harmonics which are not integer multiples of P.

Step 4: Identify the linear dynamics and nonlinear terms (if necessary) to obtain a suitable time-invariant model. If only linear dynamics are considered, the model is the best linear approximation. Otherwise, it is the best time-invariant approximation.

Step 5: Compute and plot the frequency domain indicator for determining the main source of disturbance in the system (Lataire et al. 2012). It is defined by

$$R_{\text{NTV}}(k) = \frac{6}{\pi^2} \frac{\text{E}[|W(k)|^2]}{\text{E}[W(k)\overline{W(k+1)}]}, \tag{5.10}$$

where the error signal W obtained from the difference between the actual and model outputs contains the contributions of noise and time variation. Since $R_{\text{NTV}}(k)$ cannot be computed due to the unknown expected values, it is, therefore, necessary to approximate this by the sample estimate

$$\hat{R}_{\text{NTV}}(k) = \frac{6}{\pi^2} \frac{\sum_{r=-m}^{m} \left|\hat{W}(k+r)\right|^2}{\sum_{r=-m}^{m} \hat{W}(k+r)\overline{\hat{W}(k+r+1)}}, \tag{5.11}$$

where m denotes the number of bins to the left and right of k that are being taken into account. If the indicator is significantly greater than 1, noise is likely to be the main source of disturbance whereas if it is close to 1, slowly varying parameter changes are likely to be dominant. (Methods for detecting time-varying delay can be found in Tan et al. (2015) for the single-input case and in Cham et al. (2017) for the multi-input case.)

Since the method is based on simultaneous perturbation, it inherits the many advantages that come with such an approach. In particular, the fact that only a single experiment is required means that less time is spent waiting for transient effects to decay. This is efficient in terms of both time and cost. Besides, this would minimise differences in the experimental setting caused by uncontrolled changes in the environment if several different experiments were to be conducted. Furthermore, the need to run only one experiment simplifies operational issues typically encountered in industry.

5.4.2 Application Example

The system under study is a mist reactor designed for applications in cell culture which is described in Cham et al. (2016). Further details as well as the complete experimental results are provided in Cham et al. (2017). The overall system has three inputs and two outputs, being constructed from three 1-input 1-output subsystems in

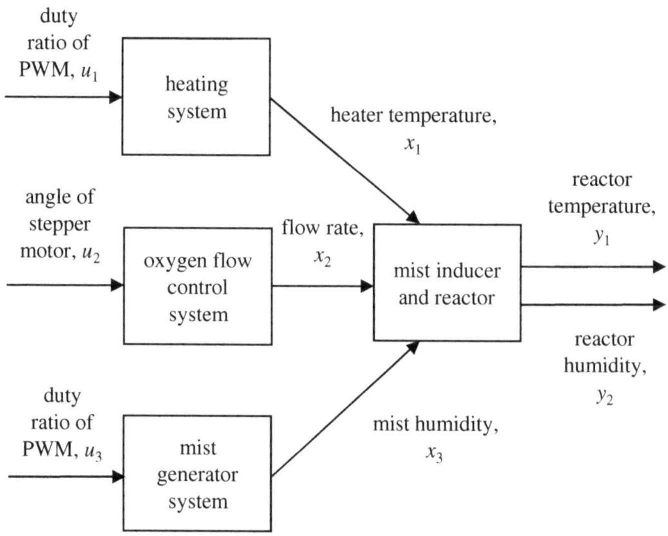

Fig. 5.9 Structure of the mist reactor system

parallel followed by a 3-input 2-output subsystem. The block diagram representation of its physical structure is illustrated in Fig. 5.9.

The main component in the heating system is a 67×85 mm^2 polymer thick film heating sheet with a power rating of 100 W. The input current to the heater is controlled by modifying the duty ratio of a pulse width modulation (PWM) signal. In the oxygen control system, the oxygen is drawn by a pump from an oxygen regulator tank. At the output of the pump, an oxygen pressure of 2 bars is sustained. A unipolar Omron stepper motor is utilised for adjusting the angle of the regulator valve. The mist generator system comprises a piezo-electric mist generator plate. Liquid with a mixture of nutrients is pumped into the piezo-electric plate. A nutrient mist spray is created as the liquid passes through the piezo-electric plate under centrifugal force. The mist generator is controlled through the duty ratio of a PWM signal that is fed into the mist injector.

The mist inducer is constructed from anodised aluminium which has very low thermal resistivity. It features an area of 60 mm × 40 mm and a height of 35 mm with a thermal resistivity of 0.0005 °C/W. Heat from the heating element is transferred to the mist in the mist inducer before entering the mist reactor. The mist reactor is a chamber with nine cylindrical compartments for cell culture, where each compartment has a height of 110 mm and a diameter of 50 mm. The temperature is measured using a K-type thermocouple placed at the centre of the reactor and the humidity is measured using a humidity sensor, with proper signal conditioning circuits in place. It was decided that it is not necessary to monitor the oxygen level as the percentage of oxygen is sufficiently high for cell culture. Data acquisition is performed through the NI PCI-6289M Series DAQ, with the timing of all the inputs and outputs being synchronised. The three intermediate signals, namely the heater temperature, flow rate and mist humidity, are not measured.

Preliminary tests in the form of input step excitations indicated that the maximum frequency of interest is 0.1 Hz. A period of $N = 600$ and a sampling frequency of 0.25 Hz were selected. The perturbation signals u_1, u_2 and u_3 were chosen as three multisine signals which are uncorrelated with one another (refer to Sect. 4.1 for the design of uncorrelated signals). The excited harmonics were chosen as follows:

$$\text{Signal A perturbing } u_1\text{: } \gamma_{\text{Signal\_A}} = \{1, 7, 13, \ldots, 235\}; \tag{5.12}$$

$$\text{Signal B perturbing } u_2\text{: } \gamma_{\text{Signal\_B}} = \{3, 9, 15, \ldots, 237\}; \tag{5.13}$$

$$\text{Signal C perturbing } u_3\text{: } \gamma_{\text{Signal\_C}} = \{5, 11, 17, \ldots, 239\}. \tag{5.14}$$

The multisines each had 40 excited harmonics with uniform DFT magnitudes. The excited harmonics were chosen to comprise of only odd harmonics. This allows the effects of even-order nonlinearities to be detected and filtered out at the system outputs. Their effects on the linear estimates can, therefore, be removed. The frequency resolution (spacing between excited harmonics) for each channel was 0.0025 Hz which was deemed to be sufficient for this application. Harmonic multiples of three were not suppressed as this would require a sparser spectrum which would lead to an increase in the experimentation time or a reduction in the frequency resolution. The decision for not suppressing the harmonic multiples of three was therefore made solely due to time considerations. In terms of the amplitude, Signals A, B and C were scaled to range from 0 to 100%, 75° to 225°, and 0 to 100%, respectively. Four periods of the outputs were collected. The first period was discarded due to the presence of transient effects. Three steady-state periods ($P = 3$) were retained for identification. The multisine test took 9600 s, which is close to 3 h. The mean values of all the signals were removed prior to identification.

The FRFs $G_{ij}(z^{-1}) = Y_i(z^{-1})/U_j(z^{-1})$ are plotted in Figs. 5.10 and 5.11 for the three individual steady-state periods and the averaged period at the excited harmonics. The input which has the largest effect on the reactor temperature is the duty ratio of the PWM signal into the heating system, which is to be expected based on common sense. In a similar way, the input which has the most significant effect on the reactor humidity is the duty ratio of the PWM signal into the mist generator system. Comparing the data for the different periods, the third period looks somewhat noisier than the first and second periods.

The 1800-point DFTs (since $N \times P = 1800$) of the outputs are plotted in Fig. 5.12. Contributions from the linear and nonlinear terms are limited to harmonics which are integer multiples of $P = 3$, since these components are assumed to be periodic with the same period as the input signals. The linear contributions corresponding to inputs u_1, u_2 and u_3 appear at harmonics $\{1, 7, 13, \ldots, 235\} \times 3 = \{3, 21, 39, \ldots, 705\}$, $\{3, 9, 15, \ldots, 237\} \times 3 = \{9, 27, 45, \ldots, 711\}$ and $\{5, 11, 17, \ldots, 239\} \times 3 = \{15, 33, 51, \ldots, 717\}$, respectively. The even-order nonlinear distortion appears at $P \times (\text{non-excited harmonics}) = \{2, 4, 6, \ldots, 300\} \times 3 = \{6, 12, 18, \ldots\}$. Disturbances such as noise and time-varying contributions fall at all the harmonics as

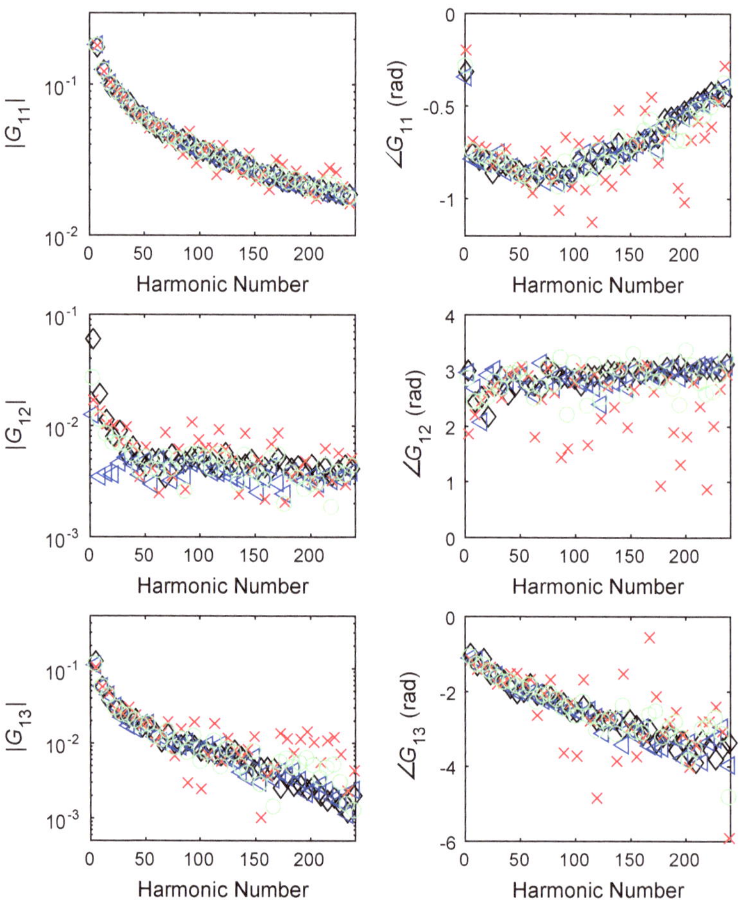

Fig. 5.10 Measured FRFs of the system for output y_1. Black diamonds: first period; blue triangles: second period; red crosses: third period; green circles: average of three individual steady-state periods (Reproduced with permission from Cham et al. © 2017 Elsevier)

these components do not have the same period as the input signals. The disturbance floor can, therefore, be represented by power at harmonics which are not integer multiples of $P = 3$, for example, at harmonics $\{1, 2, 4, 5, 7, 8, \ldots\}$.

From Fig. 5.12, it can be observed that the blue plusses and the red crosses in the top plot have almost similar magnitudes. This indicates that the effect of even-order nonlinearity in y_1 is of the same order of magnitude as the disturbance and, therefore, can be neglected. However, in the bottom plot, the blue plusses generally appear above the red crosses. There is, therefore, significant even-order nonlinear distortion appearing in y_2. It is not possible to quantify the effect of odd-order nonlinear distortion because harmonic multiples of three are not suppressed in the inputs. Their effect, if any, is absorbed into those of the linear contributions. The green diamonds

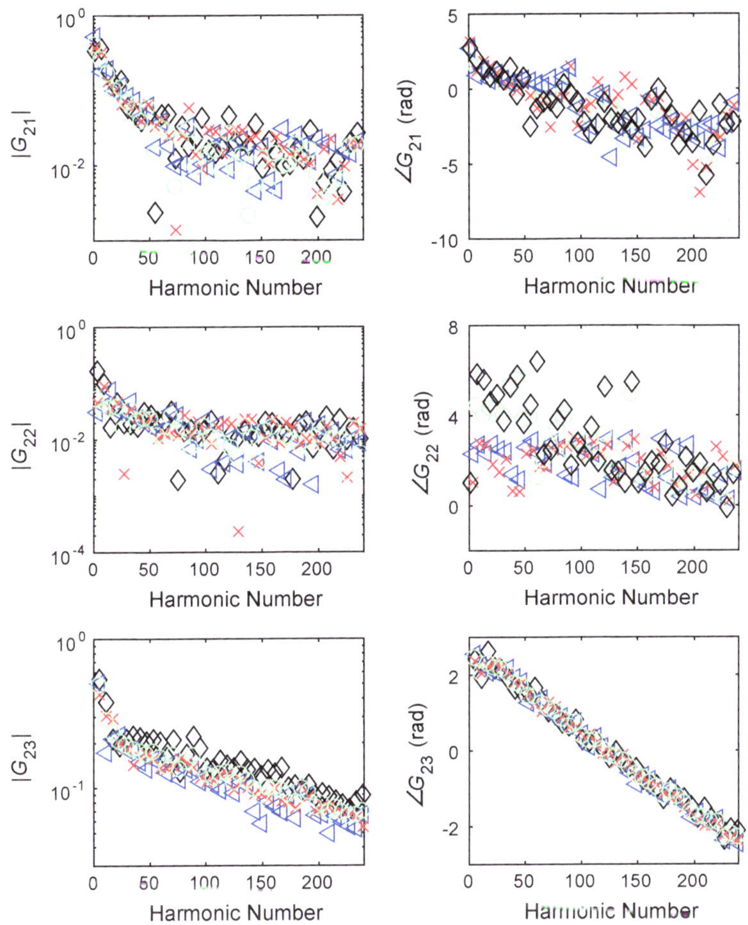

Fig. 5.11 Measured FRFs of the system for output y_2. Black diamonds: first period; blue triangles: second period; red crosses: third period; green circles: average of three individual steady-state periods (Reproduced with permission from Cham et al. © 2017 Elsevier)

are only slightly above the red crosses in the bottom plot indicating that the effect of u_2 on y_2 is small and pretty much buried in disturbance. Hence, G_{22} is expected to have a small magnitude and would be very difficult to identify accurately. (For the purpose of controller design, it is a feasible option to consider neglecting this path altogether, by setting $G_{22} = 0$, if a simpler model is sought after.)

Applying standard identification methods resulted in the best linear approximation (which is also time-invariant) given by

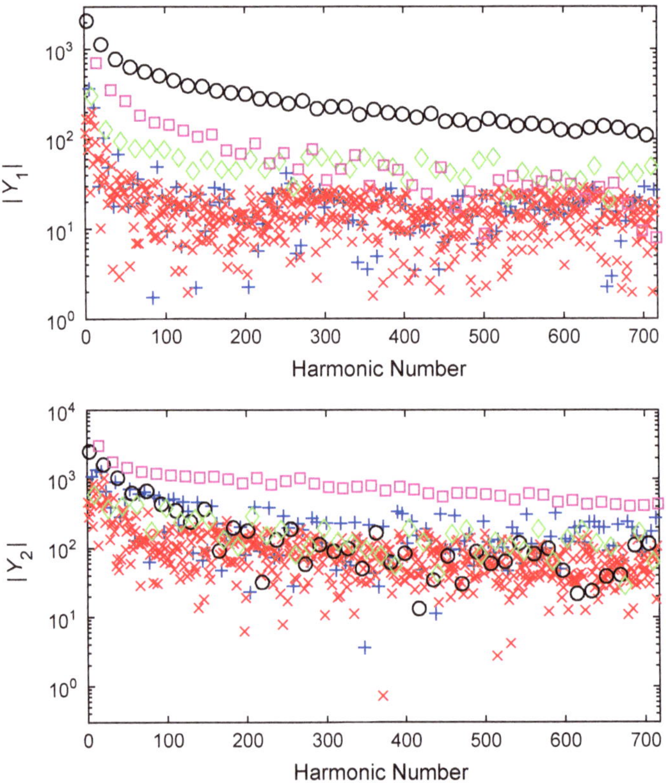

Fig. 5.12 Measured 1800-point DFT. Black circles: power at harmonics {3, 21, 39, … , 705}; green diamonds: power at harmonics {9, 27, 45, … , 711}; magenta squares: power at harmonics {15, 33, 51, … , 717}; blue plusses: power at harmonics {6, 12, 18,…}; red crosses: power at the rest of the harmonics (Reproduced with permission from Cham et al. © 2017 Elsevier)

$$\begin{bmatrix} Y_1 \\ Y_2 \end{bmatrix} = \begin{bmatrix} G_{11} & G_{12} & G_{13} \\ G_{21} & G_{22} & G_{23} \end{bmatrix} \begin{bmatrix} U_1 \\ U_2 \\ U_3 \end{bmatrix}, \tag{5.15}$$

where

$$G_{11}(z^{-1}) = \frac{0.0323 - 0.00975z^{-1} - 0.00288z^{-2} - 0.00992z^{-3}}{1 - 1.14z^{-1} + 0.257z^{-2} - 0.330z^{-3} + 0.244z^{-4}}, \tag{5.16}$$

$$G_{12}(z^{-1}) = \frac{-0.00422 + 0.00169z^{-1} - 0.00117z^{-2} + 0.00139z^{-3}}{1 - 0.752z^{-1} + 0.342z^{-2} - 0.461z^{-3}}, \tag{5.17}$$

$$G_{13}(z^{-1}) = \frac{0.00642z^{-1}}{1 - 1.31z^{-1} + 0.544z^{-2} - 0.204z^{-3}}, \tag{5.18}$$

$$G_{21}(z^{-1}) = \frac{0.00424z^{-1} - 0.0102z^{-2} - 0.00454z^{-3} - 0.00807z^{-4}}{1 - 0.979z^{-1} + 0.262z^{-2} - 0.513z^{-3} + 0.272z^{-4}}, \tag{5.19}$$

$$G_{22}(z^{-1}) = \frac{0.000362 - 0.00601z^{-1} + 0.00898z^{-2} + 0.000187z^{-3}}{1 - 0.898z^{-1} + 0.205z^{-2} - 0.147z^{-3}}, \tag{5.20}$$

$$G_{23}(z^{-1}) = \frac{-0.104z^{-2} + 0.0330z^{-3} - 0.00793z^{-4} + 0.0274z^{-5} + 0.0174z^{-6}}{1 - 1.01z^{-1} + 0.487z^{-2} - 0.524z^{-3} + 0.100z^{-4}}. \tag{5.21}$$

In line with the findings gathered from the 1800-point DFTs, an even-order nonlinearity

$$f(x) = x - 0.0177x^2 \tag{5.22}$$

was added following the linear dynamics to produce the output y_2. See Cham et al. (2017) for further details on the selection of the nonlinear structure as well as the nonlinear function. The model obtained is the best time-invariant approximation of the mist reactor.

The magnitudes of the estimated frequency domain indicator for determining the main source of disturbance, $|\hat{R}_{\mathrm{NTV}}(k)|$, are illustrated in Fig. 5.13, for $m = 10$. Computing Eq. 5.11 with different values of m will not affect the general trend in the plots except that smaller values of m will result in more fluctuations as the effect of averaging is reduced.

From Fig. 5.13, $|\hat{R}_{\mathrm{NTV}}(k)|$ for Y_1 is significantly larger than unity at frequencies below 0.02 Hz. In this range, noise is the main source of disturbance. However, above 0.02 Hz, $|\hat{R}_{\mathrm{NTV}}(k)|$ is close to unity, indicating that slowly varying parameter changes are likely to be dominant. Indeed, further analysis presented in Cham et al. (2017) shows that the channel G_{12} possesses time-varying delay. The most probable cause of this is condensation in the mist inducer.

The values of $|\hat{R}_{\mathrm{NTV}}(k)|$ for Y_2 exhibit considerable fluctuations but they are generally quite a lot larger than unity. It can thus be inferred that the effect of noise is the dominating source of disturbance. It is interesting to note that if the best linear approximation is used, the values of $|\hat{R}_{\mathrm{NTV}}(k)|$ are slightly higher than when the nonlinearity is taken into account as in the best time-invariant approximation. This is due to the effects of even-order nonlinear distortion being stochastic in nature. This example clearly shows that a carefully designed set of perturbation signals allows successful quantification of nonlinearities, noise and time variation.

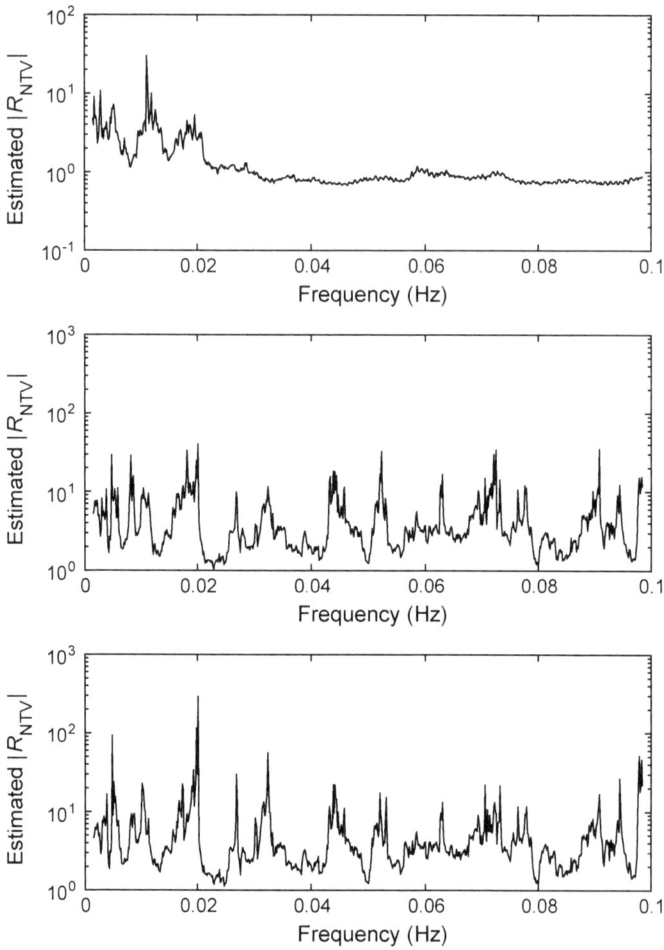

Fig. 5.13 Plots of $|\hat{R}_{\mathrm{NTV}}(k)|$. Top: Y_1; middle: Y_2 using best time-invariant approximation; bottom: Y_2 using best linear approximation

References

Ase H, Katayama T (2015) A subspace-based identification of Wiener-Hammerstein benchmark model. Control Eng Pract 44:126–137

Cham CL, Tan AH, Tan WH (2016) Design and construction of a mist reactor system. In: Proceedings of the IEEE region 10 conference (TENCON), Singapore, 22–25 November, pp 3382–3385

Cham CL, Tan AH, Tan WH (2017) Identification of a multivariable nonlinear and time-varying mist reactor system. Control Eng Pract 63:13–23

Evans C, Rees D, Jones L, Weiss M (1996) Periodic signals for measuring nonlinear Volterra kernels. IEEE Trans Instrum Meas 45:362–371

Han HT, Ma HG, Tan LN, Cao JF, Zhang JL (2014) Non-parametric identification method of Volterra kernels for nonlinear systems excited by multitone signal. Asian J Control 16:519–529

Lataire J, Louarroudi E, Pintelon R (2012) Detecting a time-varying behavior in frequency response function measurements. IEEE Trans Instrum Meas 61:2132–2143

Pintelon R, Schoukens J (2012) System identification: a frequency domain approach. Wiley, Hoboken

Pintelon R, Louarroudi E, Lataire J (2013) Detecting and quantifying the nonlinear and time-variant effects in FRF measurements using periodic excitations. IEEE Trans Instrum Meas 62:3361–3373

Schoukens M, Pintelon R, Rolain Y (2014) Identification of Wiener-Hammerstein systems by a nonparametric separation of the best linear approximation. Automatica 50:628–634

Schoukens J, Pintelon R, Rolain Y, Dobrowiecki T (2001) Frequency response function measurements in the presence of nonlinear distortions. Automatica 37:939–946

Schoukens J, Vaes M, Pintelon R (2016) Linear system identification in a nonlinear setting: nonparametric analysis of nonlinear distortions and their impact on the best linear approximation. IEEE Control Syst Mag 36:38–69

Tan AH (2007) Design of truncated maximum length ternary signals where their squared versions have uniform even harmonics. IEEE Trans Autom Control 52:957–961

Tan AH (2018) Multi-input identification using uncorrelated signals and its application to dual-stage hard disk drives. IEEE Trans Magn 54: Article 9300604

Tan AH, Godfrey KR (2002) Identification of Wiener-Hammerstein models using linear interpolation in the frequency domain (LIFRED). IEEE Trans Instrum Meas 51:509–521

Tan AH, Cham CL, Godfrey KR (2015) Comparison of three modeling approaches for a thermodynamic cooling system with time-varying delay. IEEE Trans Instrum Meas 64:3116–3123

Vanbeylen L (2015) A fractional approach to identify Wiener-Hammerstein systems. Automatica 50:903–909

Weiss M, Evans C, Rees D, Jones L (1996) Structure identification of block-oriented nonlinear systems using periodic test signal. In: Proceedings of the IEEE instrumentation and measurement technology conference, Brussels, Belgium, 4–6 June, pp 8–13

Weiss M, Evans C, Rees D (1998) Identification of nonlinear cascade systems using paired multisine signals. IEEE Trans Instrum Meas 47:332–336

Wong HK, Schoukens J, Godfrey KR (2013) Design of multilevel signals for identifying the best linear approximation of nonlinear systems. IEEE Trans Instrum Meas 62:519–524

Zhang B, Billings SA (2017) Volterra series truncation and kernel estimation of nonlinear systems in the frequency domain. Mech Syst Signal Process 84:39–57

Chapter 6
Case Study on the Identification of a Direction-Dependent Electronic Nose System

6.1 Description of System and Experimental Setup

Electronic nose research has been undertaken at the University of Warwick for many years (Gardner and Bartlett 1999), and the research team started designing, developing and building commercial electronic nose instruments in the 1990s. The Case Study described in this chapter is based on one of these electronic noses at Warwick. The development of the electronic nose has subsequently led to many applications, particularly in the food industry (Wei et al. 2015; Qiu and Wang 2017; Wojnowski et al. 2017), biochemistry (Zhao et al. 2016; Romero-Flores et al. 2017) and healthcare (Gardner and Vincent 2016; Dragonieri et al. 2016).

The gas sensor in the metal oxide semiconductor electronic nose (Gardner and Bartlett 1999) can be modelled using the adsorption-desorption reaction described by

$$\mathrm{AS} + \{\} \underset{k_b}{\overset{k_f}{\rightleftarrows}} \{\mathrm{AS}\}, \tag{6.1}$$

where AS is the adsorbed species, { } represents an empty adsorption site, {AS} represents an occupied site, k_f is the forward rate constant and k_b is the backward rate constant. A simplified relationship between the concentration of chemical C and the fractional occupancy of the adsorption sites θ is given by

$$\dot{\theta} = k_f C - \big(k_b + k_f C\big)\theta. \tag{6.2}$$

Note that the system in Eq. 6.2 is essentially bilinear, where the nonlinearity arises from the multiplicative term between input C and the state θ. However, a first-order bilinear system under a binary perturbation can be equivalently described as a direction-dependent system (Tan 2009), where the dynamics of the system depend on the direction of the output variable, whether it is increasing or decreasing. The

A. H. Tan and K. R. Godfrey, *Industrial Process Identification*, Advances in Industrial Control, https://doi.org/10.1007/978-3-030-03661-4_6

dynamics in either direction are linear, but the combined dynamics exhibit nonlinear behaviour. In particular, when C can take two values C_1 and C_2, the corresponding transfer function is given by

$$\frac{\theta(s)}{C(s)} = \begin{cases} \frac{k_f}{s+(k_b+k_f C_1)} & C = C_1 \\ \frac{k_f}{s+(k_b+k_f C_2)} & C = C_2 \end{cases}. \tag{6.3}$$

The direction-dependent dynamics can be explained by noting that the dynamics for adsorption and desorption of chemicals on the metal oxide semiconductor sensor surface are different. When the fractional occupancy θ increases, the conductance of the sensor increases. This change can be measured through a change in the output voltage.

It should be pointed out that a first-order model is a simplified model of the electronic nose. The actual dynamics are complicated, and are not yet fully understood. A recent work on the modelling of electronic nose based on metal oxide semiconductor is found in Guo et al. (2015).

In the identification tests which are described in Tan and Godfrey (2004), the input C was controlled by adjusting the position of the valves through which chemical odours can reach the metal oxide semiconductor sensor. Acetone was used as the positive input, and air as the negative input. The positive input was achieved by opening the valve which connected the electronic nose sensor to the container containing acetone and closing the valve to the container containing air. The negative input was achieved by opening the valve which connected the sensor to the container containing air and closing the valve to the container containing acetone.

It is interesting to note that the number of signal levels was limited by the number of containers used since the valves were programmed to be either fully open or fully closed. The physical construction of the system allowed the use of up to four signal levels as there were four containers. However, in the experiments conducted in this study, only two containers were utilised. Data acquisition was carried out using the LabVIEW software. The measured output is the voltage across the sensor, which is inversely proportional to the conductance.

6.2 Detection of Nonlinear Distortion

6.2.1 Detection Through Step Tests

Step tests were first conducted, using a sampling interval of 1 s. Results are plotted in Fig. 6.1 where it can be observed that the response in the upward direction is slower than that in the downward direction, thus confirming the presence of direction-dependent dynamics. In particular, the time constants of first-order models in the

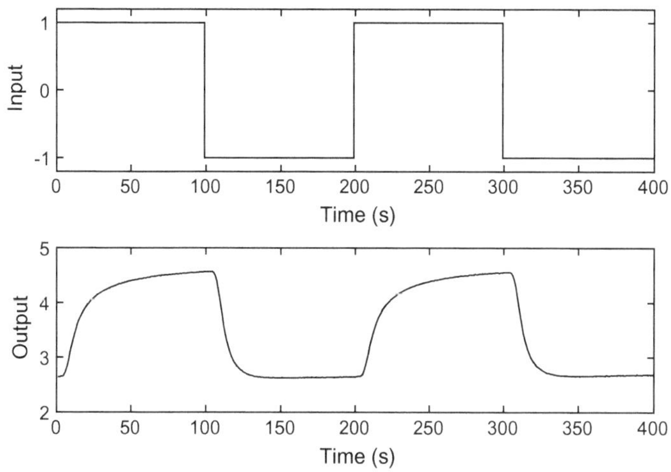

Fig. 6.1 Step tests for the electronic nose system

upward and downward directions were found to be 18.49 and 5.54 s, respectively. The system also has a dead time of approximately 3 s, which arises because the chemical takes some time after a valve is opened to make contact with the sensor. From Fig. 6.1, it can also be observed that the response is only approximately first order. The initial part of the downward dynamics only partly resembles an exponential.

6.2.2 Detection Through Crosscorrelation Function

The system was then perturbed in two separate experiments, each using an MLB signal. The first MLB signal, Signal 1, has a characteristic equation given by $D^6 \oplus_2 D^5 \oplus_2 D^4 \oplus_2 D \oplus_2 1 = 0$ while the second MLB signal, Signal 2, has characteristic equation given by $D^6 \oplus_2 D^5 \oplus_2 D^3 \oplus_2 D^2 \oplus_2 1 = 0$. Both signals have period $N = 63$. Signal 1 was used as training signal, whereas Signal 2 was reserved as validation signal. The output was sampled at 1 s while the clock pulse interval for the input was set to 10 s. The mean value of the output and a dead time of 3 s were also removed prior to identification. The input–output crosscorrelation function using Signal 1 is plotted against the delay in Fig. 6.2.

From Fig. 6.2, the crosscorrelation function is observed to have some coherent peaks besides the main peak starting at delay = 0. In particular, there are negative peaks starting at delay positions 39, 15 and 11. (For Signal 2, these occur starting at delay positions 8, 16 and 53. See Tan and Godfrey (2004) for the corresponding plot.) The presence of the peaks can be explained based on the following theoretical analysis.

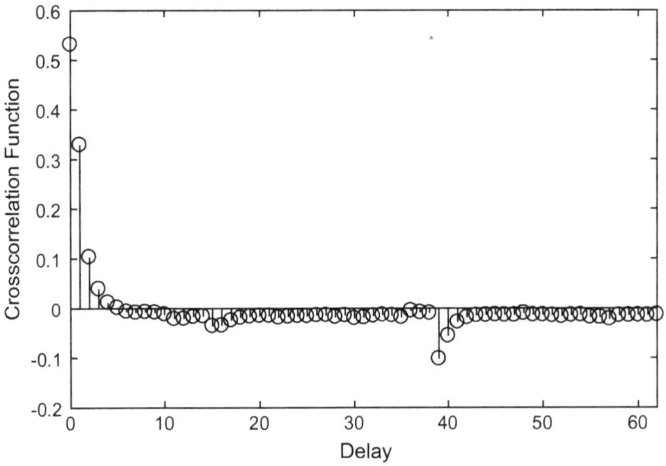

Fig. 6.2 Normalised input–output crosscorrelation function using Signal 1. The delay is normalised with respect to the clock pulse interval

Consider a first-order direction-dependent system in discrete time, with K_U and T_U denoting the gain and time constant in the positive direction, and K_D and T_D denoting the gain and time constant in the negative direction. The output $y(i)$ in response to an MLB signal $u(i)$ with levels ± 1 is given by Tan and Godfrey (2001)

$$\begin{aligned} y(i) &= (a + bu(i))y(i-1) + (1 - a - bu(i))(Au(i) + F) \\ &= ay(i-1) + bu(i)y(i-1) + (A - aA - bF)u(i) - bA + (1-a)F \\ &= (a + bu(i))\left(y(i-1) + \left(\frac{A - aA - bF}{b}\right)\right) - bA + (1-a)F \\ &\quad - \left(\frac{A - aA - bF}{b}\right)a \end{aligned} \tag{6.4}$$

where

$$a = 0.5(\exp(-T/T_U) + \exp(-T/T_D)), \tag{6.5}$$

$$b = 0.5(\exp(-T/T_U) - \exp(-T/T_D)), \tag{6.6}$$

$$A = \frac{K_U + K_D}{2}, \tag{6.7}$$

$$F = \frac{K_U - K_D}{2} \tag{6.8}$$

and T is the clock pulse interval. Further manipulation of Eq. 6.4 leads to

$$y_t = A\left(\left(\frac{(1-a)^2}{b} - b\right)\sum_{p=0}^{v}\prod_{l=0}^{p}(a + bu(i-l)) - \left(\frac{1-a}{b}\right)\right) + F. \tag{6.9}$$

The right-hand side of Eq. 6.9 can be split into constant, first-order and higher order terms (Tan and Godfrey 2001):

$$\text{Constant term: } y_0 = -\frac{Ab}{1-a} + F, \tag{6.10}$$

$$\text{First-order term: } y_1 = AK\frac{b}{1-a}\sum_{k=0}^{\infty} a^k u(i-k), \tag{6.11}$$

$$\text{Second-order terms: } y_2 = AK\frac{b^2}{1-a}\sum_{k=0}^{\infty} a^k u(i-k-1)\sum_{m=0}^{k} u(i-m), \tag{6.12}$$

$$\begin{aligned}\text{Third-order terms: } y_3 = {} & AK\frac{b^3}{1-a} \\ & \times \sum_{k=0}^{\infty} a^k u(i-k-2)\sum_{m=0}^{k} u(i-m-1)\sum_{n=0}^{m} u(i-n),\end{aligned} \tag{6.13}$$

where $K = \frac{(1-a)^2}{b} - b$.

Consider the second-order terms. Expanding Eq. 6.12 for k up to 3 and reordering some terms give

$$\begin{aligned} y_2 = {} & AK\frac{b^2}{1-a} \\ & \times ([a^0u(i)u(i-1) + a^1u(i-1)u(i-2) + a^2u(i-2)u(i-3) \\ & + a^3u(i-3)u(i-4) + \cdots] \\ & + [a^1u(i)u(i-2) + a^2u(i-1)u(i-3) + a^3u(i-2)u(i-4) + \cdots] \\ & + [a^2u(i)u(i-3) + a^3u(i-1)u(i-4) + \cdots] \\ & + [a^3u(i)u(i-4) + \cdots] + \cdots). \end{aligned} \tag{6.14}$$

From the shift-and-multiply property of the MLB signals (Eq. 2.6) where $u(i-\alpha)u(i-\beta) = -u(i-\gamma)$, the second-order terms can be replaced by single-order terms. This causes negative peaks in the crosscorrelation function. The starting positions of these peaks are shown in Table 6.1 and depend on the exact MLB signal used. The magnitude of these peaks decreases down Table 6.1, in accordance with the decay at a rate of a in Eq. 6.14.

Similarly, for the third-order terms, expanding Eq. 6.13 for k up to 2 results in

Table 6.1 Starting positions of second-order peaks in the crosscorrelation function

β	γ for Signal 1	γ for Signal 2
1	39	8
2	15	16
3	11	53
4	30	32
5	28	38
6	22	43
7	17	62
8	60	1
9	45	45
10	56	13

These are given by the values of γ which satisfy Eq. 2.6, $u(i-\alpha)u(i-\beta) = -u(i-\gamma)$, with $\alpha = 0$

$$\begin{aligned}
y_3 = AK\frac{b^3}{1-a}
&\times ([a^0u(i)u(i-1)u(i-2) + a^1u(i-1)u(i-2)u(i-3) \\
&+ a^2u(i-2)u(i-3)u(i-4) + \cdots] \\
&+ [a^1u(i)u(i-1)u(i-3) + a^2u(i-1)u(i-2)u(i-4) + \cdots] \\
&+ [a^1u(i)u(i-2)u(i-3) + a^2u(i-1)u(i-3)u(i-4) + \cdots] \\
&+ [a^2u(i)u(i-1)u(i-4) + \cdots] \\
&+ [a^2u(i)u(i-2)u(i-4) + \cdots] \\
&+ [a^2u(i)u(i-3)u(i-4) + \cdots] + \cdots).
\end{aligned} \tag{6.15}$$

According to the shift-and-multiply property of the MLB (Eq. 2.9), $u(i-\alpha)u(i-\beta)u(i-\chi) = u(i-\gamma)$. The starting positions of these peaks are shown in Table 6.2. These peaks are positive and are much smaller in magnitude than the second-order peaks. In fact, they are mostly buried in noise. Fourth-order and higher order peaks become increasingly small and can be considered negligible. These results illustrate the effectiveness of the MLB signal in the detection of nonlinear distortion through crosscorrelation function.

Experiments were also carried out using inverse-repeat versions of Signal 1 and Signal 2. The crosscorrelation function is plotted in Fig. 6.3 for Signal 1. The second-order peaks do not appear in the crosscorrelation function. The third-order peaks remain very small and are not really noticeable. While the nonlinear distortion cannot be easily detected when inverse-repeat MLB signals are applied, these signals can result in better linear estimates compared to the original MLB signals since the effects of even-order nonlinear distortion can be completely eliminated.

The delays γ in Tables 6.1 and 6.2 can be found by the following codes in MATLAB®. (MATLAB® is a registered product of The MathWorks, Inc.) This

Table 6.2 Starting positions of third-order peaks in the crosscorrelation function

β	χ	γ for Signal 1	γ for Signal 2
1	2	35	45
1	3	57	41
2	3	47	13
1	4	44	36
2	4	7	27
3	4	21	51
1	5	19	58
2	5	61	56
3	5	27	34
4	5	29	23

These are given by the values of γ which satisfy Eq. 2.9, $u(i-\alpha)u(i-\beta)u(i-\chi) = u(i-\gamma)$, with $\alpha = 0$

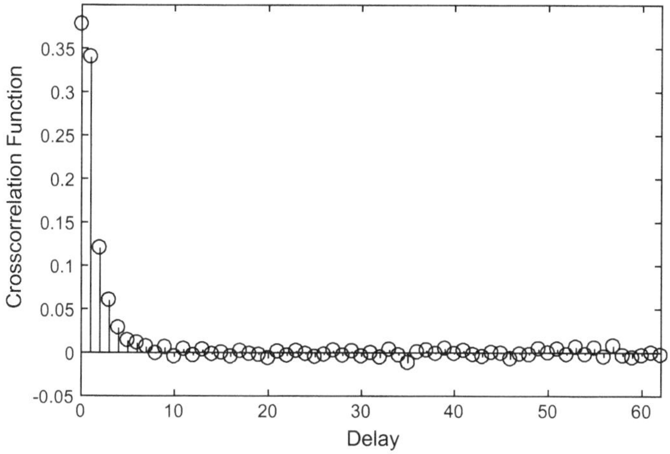

Fig. 6.3 Normalised input output crosscorrelation function using inverse-repeat version of Signal 1. The delay is normalised with respect to the clock pulse interval

applies for any MLB signal. The codes make use of the negative peak in the crosscorrelation between $u(i)u(i-\beta)$ and $u(i)$ to determine the value of γ for the second-order terms, and the positive peak in the crosscorrelation between $u(i)u(i-\beta)u(i-\chi)$ and $u(i)$ to determine the value of γ for the third-order terms. The crosscorrelation function between $u(i)u(i-1)$ and $u(i)$ as well as that between $u(i)u(i-1)u(i-2)$ and $u(i)$ for Signal 1 are shown in Figs. 6.4 and 6.5, respectively.

```
%u is the MLB signal
N=length(u);
%udx is the signal u delayed by x
ud1(1)=u(N);ud1(2:N)=u(1:N-1);ud1=ud1';
ud2(1)=ud1(N);ud2(2:N)=ud1(1:N-1);ud2=ud2';
```

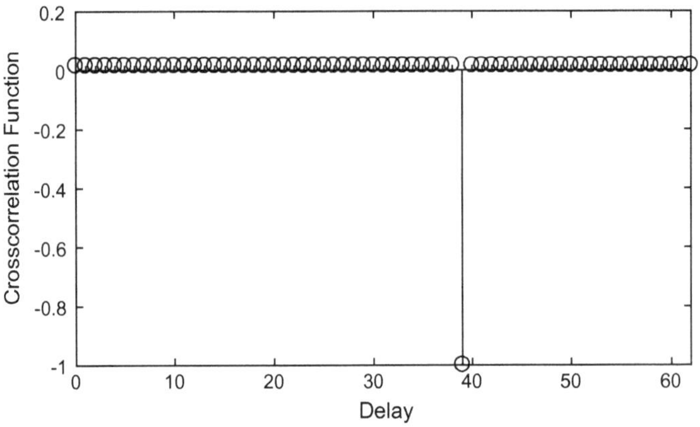

Fig. 6.4 Crosscorrelation function between $u(i)u(i-1)$ and $u(i)$ for Signal 1

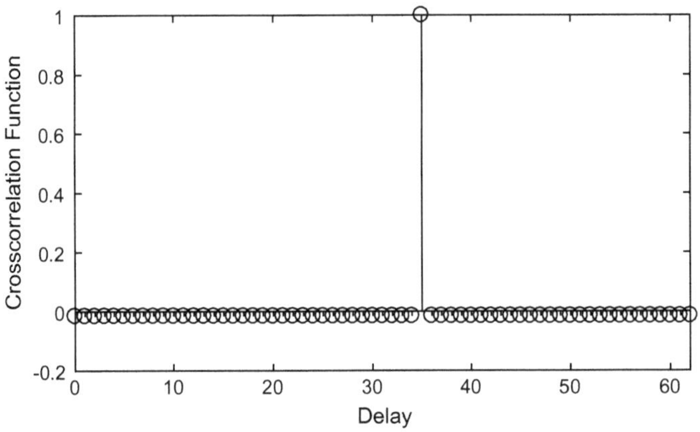

Fig. 6.5 Crosscorrelation function between $u(i)u(i-1)u(i-2)$ and $u(i)$ for Signal 1

```
ud3(1)=ud2(N);ud3(2:N)=ud2(1:N-1);ud3=ud3';
ud4(1)=ud3(N);ud4(2:N)=ud3(1:N-1);ud4=ud4';
ud5(1)=ud4(N);ud5(2:N)=ud4(1:N-1);ud5=ud5';
ud6(1)=ud5(N);ud6(2:N)=ud5(1:N-1);ud6=ud6';
ud7(1)=ud6(N);ud7(2:N)=ud6(1:N-1);ud7=ud7';
ud8(1)=ud7(N);ud8(2:N)=ud7(1:N-1);ud8=ud8';
ud9(1)=ud8(N);ud9(2:N)=ud8(1:N-1);ud9=ud9';
ud10(1)=ud9(N);ud10(2:N)=ud9(1:N-1);ud10=ud10';

%calculate second order shifts
%locate the negative peak through crosscorrelation
R=real(ifft(conj(fft(u)).*fft(u.*ud1)))/N;
[i,j]=find(R<0);i-1
R=real(ifft(conj(fft(u)).*fft(u.*ud2)))/N;
[i,j]=find(R<0);i-1
R=real(ifft(conj(fft(u)).*fft(u.*ud3)))/N;
```

```
[i,j]=find(R<0);i-1
R=real(ifft(conj(fft(u)).*fft(u.*ud4)))/N;
[i,j]=find(R<0);i-1
R=real(ifft(conj(fft(u)).*fft(u.*ud5)))/N;
[i,j]=find(R<0);i-1
R=real(ifft(conj(fft(u)).*fft(u.*ud6)))/N;
[i,j]=find(R<0);i-1
R=real(ifft(conj(fft(u)).*fft(u.*ud7)))/N;
[i,j]=find(R<0);i-1
R=real(ifft(conj(fft(u)).*fft(u.*ud8)))/N;
[i,j]=find(R<0);i-1
R=real(ifft(conj(fft(u)).*fft(u.*ud9)))/N;
[i,j]=find(R<0);i-1
R=real(ifft(conj(fft(u)).*fft(u.*ud10)))/N;
[i,j]=find(R<0);i-1

%calculate third order shifts
%locate the positive peak through crosscorrelation
R=real(ifft(conj(fft(u)).*fft(u.*ud1.*ud2)))/N;
[i,j]=find(R>0);i-1
R=real(ifft(conj(fft(u)).*fft(u.*ud1.*ud3)))/N;
[i,j]=find(R>0);i-1
R=real(ifft(conj(fft(u)).*fft(u.*ud2.*ud3)))/N;
[i,j]=find(R>0);i-1
R=real(ifft(conj(fft(u)).*fft(u.*ud1.*ud4)))/N;
[i,j]=find(R>0);i-1
R=real(ifft(conj(fft(u)).*fft(u.*ud2.*ud4)))/N;
[i,j]=find(R>0);i-1
R=real(ifft(conj(fft(u)).*fft(u.*ud3.*ud4)))/N;
[i,j]=find(R>0);i-1
R=real(ifft(conj(fft(u)).*fft(u.*ud1.*ud5)))/N;
[i,j]=find(R>0);i-1
R=real(ifft(conj(fft(u)).*fft(u.*ud2.*ud5)))/N;
[i,j]=find(R>0);i-1
R=real(ifft(conj(fft(u)).*fft(u.*ud3.*ud5)))/N;
[i,j]=find(R>0);i-1
R=real(ifft(conj(fft(u)).*fft(u.*ud4.*ud5)))/N;
[i,j]=find(R>0);i-1
```

6.2.3 Detection Through Output Spectrum

The use of inverse-repeat MLB signals has an additional advantage such that the nonlinear distortion can be detected in the output DFT. This is because the power at the even harmonics has been suppressed at the input. The output DFT using the inverse-repeat version of Signal 1 is shown in Fig. 6.6, where the power at the even harmonics at the output can be attributed to the effects of even-order nonlinear distortion (and some noise).

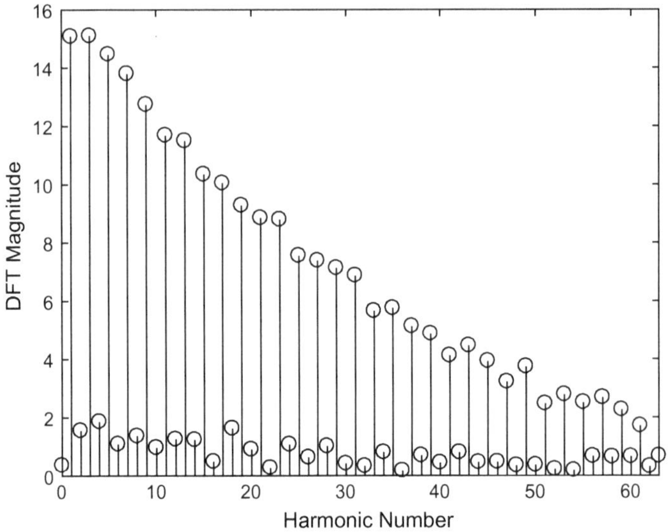

Fig. 6.6 DFT magnitude of the output signal using inverse-repeat version of Signal 1

6.3 Identification of Linear Dynamics

The value of a in Eq. 6.5 is directly related to the best linear approximation (also referred to as the combined linear dynamics in some literature) of the system. In particular, the time constant corresponding to the best linear approximation is given by

$$T_C = -\frac{T}{\ln(a)}. \tag{6.16}$$

The equivalent impulse response $w_C(iT)$ of the best linear approximation (Barker et al. 2003) is given by

$$w_C(iT) = \begin{cases} (Ac/(1-a))a^i & i \geq 0 \\ 0 & i < 0 \end{cases}, \tag{6.17}$$

where

$$c = (1 - \exp(-T/T_U))(1 - \exp(-T/T_D)). \tag{6.18}$$

Based on the step response tests, T_U =18.49 s and T_D =5.54 s. From Eqs. 6.5 and 6.16, a =0.3734 giving the theoretical value of T_C =10.15 s. However, the theoretical combined time constant may not be very accurate as this was calculated with the assumption that the process was strictly first order.

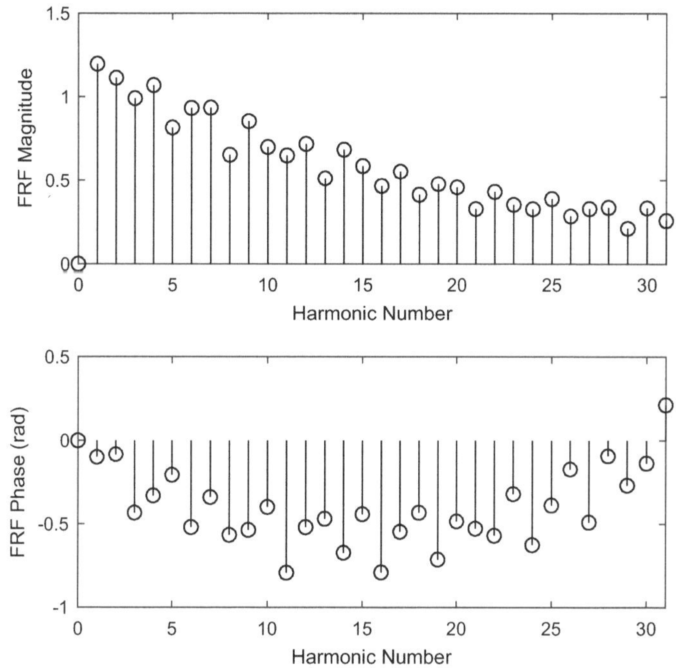

Fig. 6.7 FRF measured using Signal 1

As for the gains, an approximation can be made that $K_U = K_D = 1.25$ based on the FRF plot in Fig. 6.7. The FRF magnitude was slightly extrapolated to cover harmonic 0. The gains were assumed to be equal in both increasing and decreasing directions of the output as there is no further information to distinguish between K_U and K_D since the input is binary and mean value of neither the input signal nor output signal was taken into account in the identification.

The best linear approximation was then estimated using both Signal 1 and its inverse-repeat version. Four different models/algorithms were applied, namely the autoregressive with exogenous input (ARX), autoregressive moving average with exogenous input (ARMAX), state-space and maximum likelihood implemented using ELiS in the Frequency Domain System Identification Toolbox (Kollár 1994). The first three of these are based on time domain estimation whereas ELiS uses frequency domain estimation. Models with 1 pole as well as models with 1 zero and 2 poles were tested. The estimated combined time constants are tabulated in Table 6.3. Higher confidence should be placed on the results obtained using the inverse-repeat signal as the effects of even-order nonlinearities were eliminated in this case.

Table 6.3 Estimated combined time constant T_C

Model/algorithm	Signal 1 Order 0/1, T_C	Signal 1 Order 1/2, T_C	Inverse-repeat of Signal 1 Order 0/1, T_C	Inverse-repeat of Signal 1 Order 1/2, T_C
ARX	13.39	10.51	18.13	12.06
ARMAX	13.39	10.51	18.17	12.02
State-space	11.72	9.50, 55.38	11.27	5.08, 17.12
ELiS	12.90	12.67, 40.35	15.19	15.19

For models with two poles, only the positive poles were considered

6.4 Identification of Wiener Model

The Wiener model was chosen to approximate direction-dependent dynamics as the crosscorrelation function is very similar to that for a direction-dependent system. Furthermore, block-oriented models have the advantage of having simple structures.

The constant component R_{uy0}, linear component R_{uy1}, quadratic component R_{uy2} and cubic component R_{uy3} of the crosscorrelation function for a first-order direction-dependent system perturbed using an MLB input are given by Barker et al. (2003)

$$R_{uy0}(iT) = -K_D/N, \tag{6.19}$$

$$R_{uy1}(iT) = ((N+1)/N)w_C(iT), \tag{6.20}$$

$$R_{uy2}(iT) = -\frac{N+1}{N}b\sum_{r=1}^{N-1} a^{r-1}w_C((i-\gamma)T), \tag{6.21}$$

where γ is the position of delay satisfying $u(i)u(i-r) = -u(i-\gamma)$, and

$$R_{uy3}(iT) = \frac{N+1}{N}b^2\sum_{q=1}^{N-2}\sum_{r=1}^{N-1-q} a^{q+r-2}w_C((i-\gamma)T), \tag{6.22}$$

where γ is the position of delay satisfying $u(i)u(i-q)u(i-q-r) = u(i-\gamma)$. From Eq. 6.21, the quadratic component is the sum of scaled and shifted replicas of the impulse response $w_C(iT)$. The scale of the shifted replicas is proportional to a^r, and therefore decreases as r increases. From Eq. 6.22, the cubic component is the sum of scaled and shifted replicas of w_C (iT). The scale of the shifted replicas is proportional to a^{q+r}, and decreases as $q+r$ increases.

If the system input is an inverse-repeat MLB signal, then the crosscorrelation function is also inverse-repeat with no even-order components. As shown by Tan and Godfrey (2001), the linear and cubic components in the first half-period of the crosscorrelation function are identical to those in Eqs. 6.20 and 6.22, except for the addition of very small oscillatory biases.

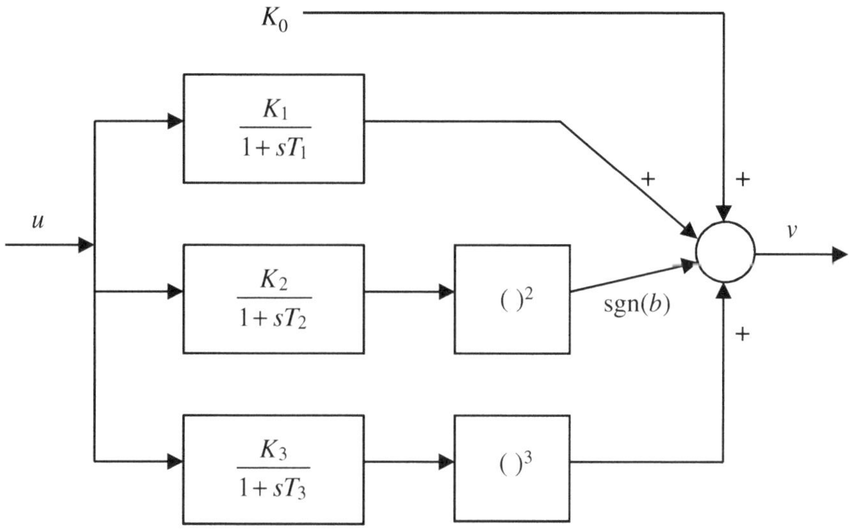

Fig. 6.8 Wiener model of a first-order direction-dependent system (Reproduced with permission from Barker et al. © 2003 Elsevier)

The Wiener model used for matching the direction-dependent system has four paths, corresponding to the four most significant components of the crosscorrelation function. The block diagram of the model is shown in Fig. 6.8. The contribution of the quadratic path should be either added or subtracted to form the total output according to the sign of b in Eq. 6.6. In particular, sgn(b) = 1 if $T_U > T_D$ and sgn(b) = -1 if $T_U < T_D$.

For this model, the discrete periodic crosscorrelation function $\phi_{uv}(iT)$ between the model input $u(t)$ and the steady-state model output $v(t)$ is the sum of four components, each corresponding to a path in the model. Analytical expressions for these components have been obtained by Barker and Obidegwu (1973). The expression for the mth order path involves the parameter

$$a_m = \exp(-T/T_m) \tag{6.23}$$

which defines two impulse responses associated with the path dynamics. The first impulse response corresponds to the linear block in the path, with transfer function $W_m(s) = K_m/(1 + sT_m)$, that is

$$w_m(iT) = \begin{cases} K_m(1 - a_m)a_m^i & i \geq 0 \\ 0 & i < 0 \end{cases}. \tag{6.24}$$

The second impulse response is that of linear dynamics with transfer function $W'_m(s) = K_m^m/(1 + s(T_m/m))$, that is

$$w'_m(iT) = \begin{cases} K_m^m(1 - a_m^m)a_m^{mi} & i \geq 0 \\ 0 & i < 0 \end{cases}. \tag{6.25}$$

The constant component R_{uv0}, linear component R_{uv1}, quadratic component R_{uv2} and cubic component R_{uv3} of the model crosscorrelation function are then given by Barker et al. (2003)

$$R_{uv0}(iT) = K_0/N, \tag{6.26}$$

$$R_{uv1}(iT) = ((N + 1)/N)w_1(iT) - K_1/N, \tag{6.27}$$

$$R_{uv2}(iT) = -\mathrm{sgn}(b)\frac{N + 1}{N}2\frac{1 - a_2}{1 + a_2}\sum_{r=1}^{N-1} a_2^r w'_2((i - \gamma)T) + \mathrm{sgn}(b)\frac{K_2^2}{N}, \tag{6.28}$$

$$R_{uv3}(iT) = \frac{N + 1}{N}6\frac{(1 - a_3)^3}{1 - a_3^3}\sum_{q=1}^{N-2}\sum_{r=1}^{N-1-q} a_3^{2q+r} w'_3((i - \gamma)T)$$
$$+ \frac{N + 1}{N}3K_3^2\frac{1 - a_3}{1 + a_3}w_3(iT) - \frac{N + 1}{N}2\frac{(1 - a_3)^3}{1 - a_3^3}w'_3(iT) - \frac{K_3^3}{N}. \tag{6.29}$$

From Eq. 6.28, the quadratic component is the sum of scaled and shifted replicas of the impulse response $w'_2(iT)$ and a bias. The scale of the shifted replicas is proportional to a_2^r, and therefore decreases as r increases. From Eq. 6.29, the cubic component is the sum of scaled and shifted replicas of the impulse response $w'_3(iT)$, a scaled replica of the impulse response $w_3(iT)$ and a bias. The scale of the shifted replicas is proportional to a_3^{2q+r}, and therefore decreases as $2q + r$ increases. The quadratic and cubic components consist of terms which will result in peaks similar to those observed earlier for the direction-dependent system.

If the model input is an inverse-repeat MLB signal, then the crosscorrelation function is also inverse-repeat with no even-order components. As shown by Barker and Obidegwu (1973), the linear and cubic components in the first half-period of the crosscorrelation function are identical to those in Eqs. 6.27 and 6.29, except that the biases are replaced by very small oscillatory biases.

The Wiener model parameters were estimated by matching the outputs of the actual electronic nose system and the Wiener model and retaining the parameters which resulted in the minimum mean square error (MSE) such that

$$\text{optimised parameters} = \arg \min_{K_0,K_1,T_1,K_2,T_2,K_3,T_3} \left[\sum_{i=0}^{N-1} [y(iT) - v(iT)]^2\right]. \tag{6.30}$$

The optimisation was performed using nonlinear multivariable minimisation. Since b in Eq. 6.6 is 0.2089, $\mathrm{sgn}(b) = 1$ in the quadratic path of Fig. 6.8.

The parameters estimated using Signal 1 were $K_0 = -0.08$, $K_1 = 1.47$, $K_2 = 0.66$, $K_3 = -0.86$, $T_1 = 19.34$ s, $T_2 = 25.82$ s and $T_3 = 30.00$ s (Tan and Godfrey 2004). The input, actual output, model output and error signals are plotted in Fig. 6.9 for the training set using Signal 1. These are plotted in Fig. 6.10 for the validation set using Signal 2. An MSE of 0.0687 was evaluated based on the validation set. From Figs. 6.9 and 6.10, it can be seen that the Wiener model successfully captured the trend in the output of the electronic nose.

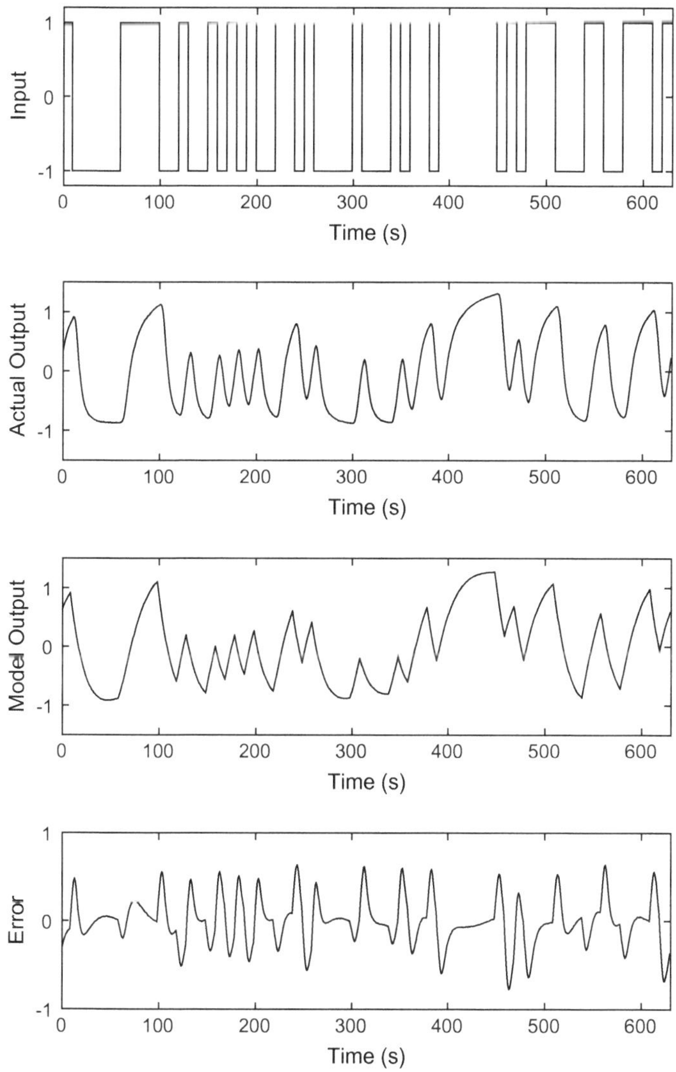

Fig. 6.9 Input, actual output, model output and error signals using Signal 1

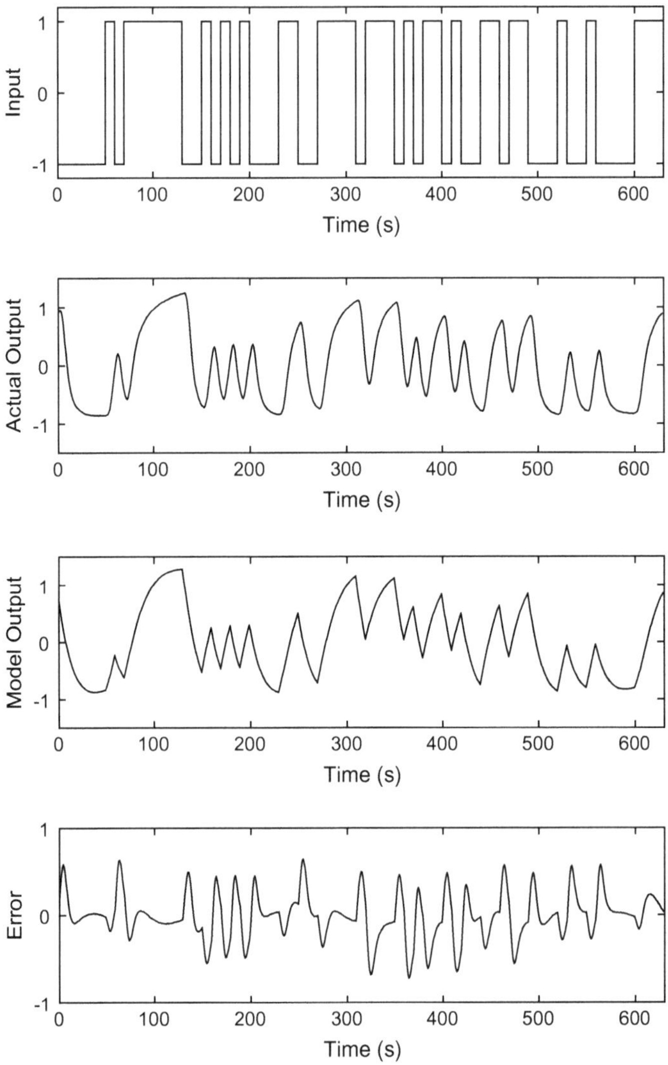

Fig. 6.10 Input, actual output, model output and error signals using Signal 2

The modelling was repeated for the perturbation using inverse-repeat signals. Since the signals have even harmonics suppressed, the odd- and even-order terms can be separated at the output. These are formed using

$$y_{\text{odd}}(iT) = \begin{cases} \frac{1}{2}\left[y(iT) - y\left(\left(i + \frac{N}{2}\right)T\right)\right] & 0 \le i < \frac{N}{2} \\ -y_{\text{odd}}\left(\left(i - \frac{N}{2}\right)T\right) & \frac{N}{2} \le i < N \end{cases}, \tag{6.31}$$

$$y_{\text{even}}(iT) = \begin{cases} \frac{1}{2}\left[y(iT) + y\left(\left(i + \frac{N}{2}\right)T\right)\right] & 0 \le i < \frac{N}{2} \\ y_{\text{even}}\left(\left(i - \frac{N}{2}\right)T\right) & \frac{N}{2} \le i < N \end{cases}. \tag{6.32}$$

The optimisation was carried out separately for these components such that

$$\text{optimised odd parameters} = \arg \min_{K_1, T_1, K_3, T_3} \left[\sum_{i=0}^{N-1} [y_{\text{odd}}(iT) - v_{\text{odd}}(iT)]^2 \right], \tag{6.33}$$

$$\text{optimised even parameters} = \arg \min_{K_0, K_2, T_2} \left[\sum_{i=0}^{N-1} [y_{\text{even}}(iT) - v_{\text{even}}(iT)]^2 \right]. \tag{6.34}$$

The parameters estimated using the inverse-repeat of Signal 1 were $K_0 = -0.03$, $K_1 = 1.20$, $K_2 = 0.56$, $K_3 = -0.66$, $T_1 = 20.21$ s, $T_2 = 43.04$ s and $T_3 = 23.27$ s (Tan and Godfrey 2004). The input, actual output, model output and error signals are plotted in Figs. 6.11 and 6.12 separately for the odd and even components for the training set. It is interesting to see that both the odd and even components of the Wiener model effectively captured the pattern in the electronic nose output. The MSE was evaluated at the combined output based on the validation set. This gave a value of 0.0427 which was considerably smaller than the one obtained earlier, the reason being that the separate fitting of the odd and even components reduced the optimisation problem into two smaller problems which can be solved more effectively. The performance of the estimated Wiener model in describing the actual system is excellent as can be observed from Fig. 6.13.

This Case Study has shown the effectiveness of the MLB signal and its corresponding inverse-repeat version in the identification of the electronic nose system for which hardware constraints faced during the experiments precluded the application of signals with more than two levels.

Further work on the modelling of the electronic nose using neural networks and piecewise linear models can be found in Tan and Godfrey (2004) and Rosenqvist et al. (2006), respectively.

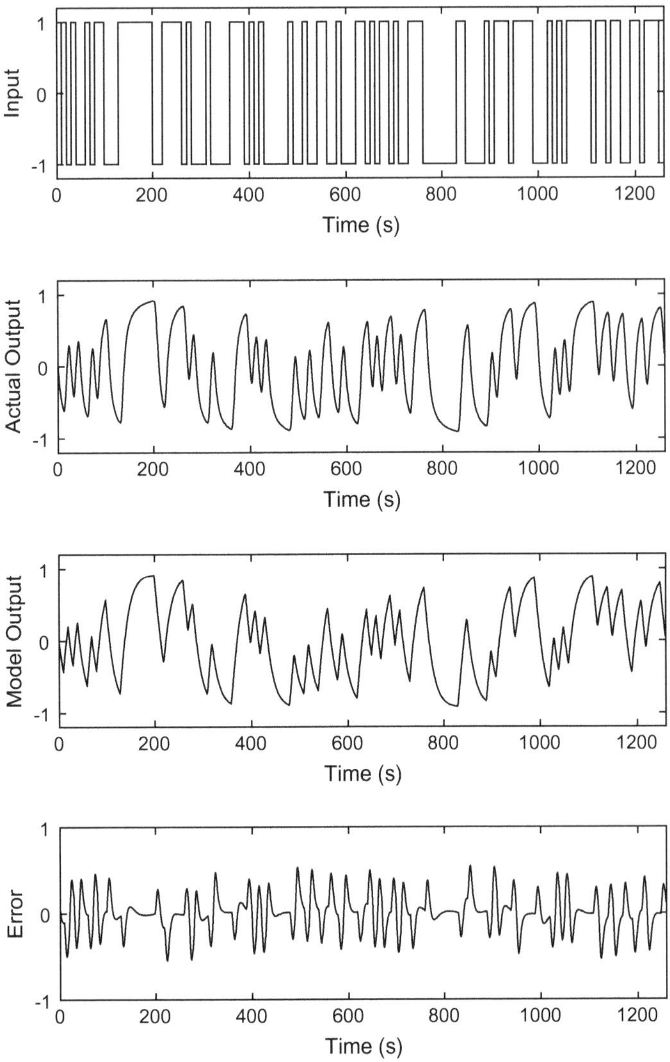

Fig. 6.11 Input, actual output, model output and error signals for the odd components using the inverse-repeat version of Signal 1

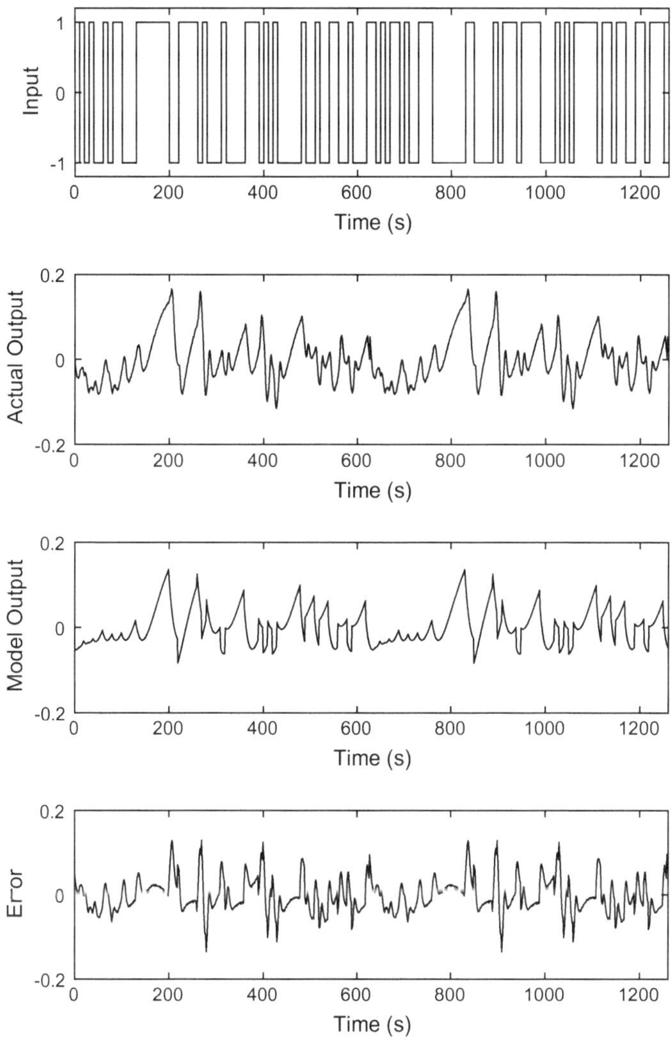

Fig. 6.12 Input, actual output, model output and error signals for the even components using the inverse-repeat version of Signal 1

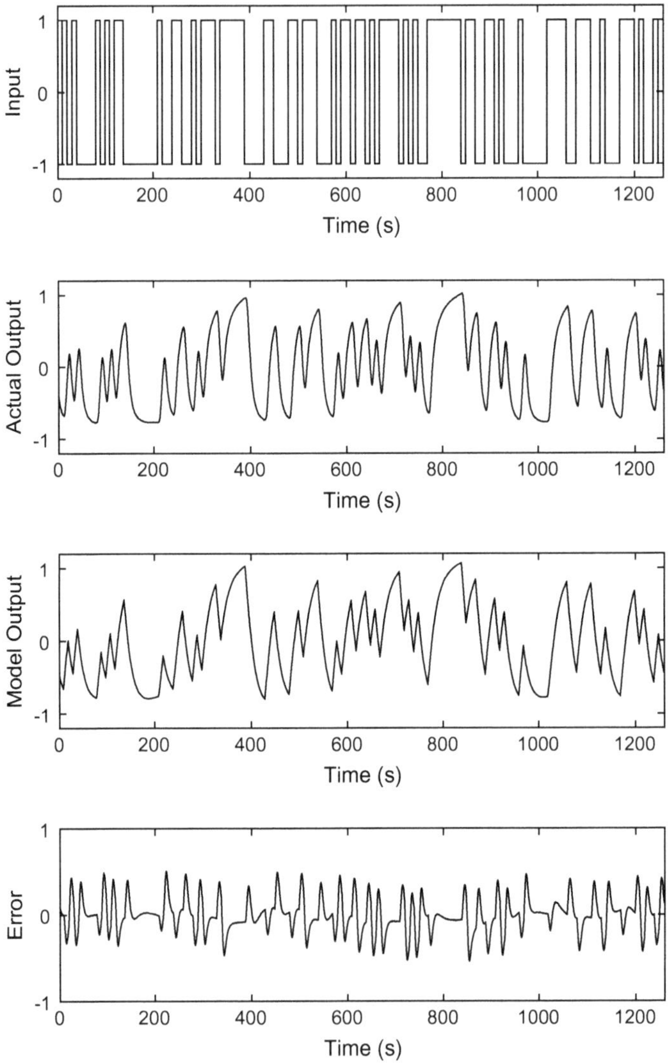

Fig. 6.13 Input, actual output, model output and error signals using the inverse-repeat version of Signal 2

References

Barker HA, Obidegwu SN (1973) Effects of nonlinearities on the measurement of weighting functions by crosscorrelation using pseudorandom signals. IEE Proc Control and Science 120:1293–1300

Barker HA, Tan AH, Godfrey KR (2003) Wiener models of direction-dependent dynamic systems. Automatica 39:127–133

Dragonieri S, Quaranta VN, Carratu P, Ranieri T, Marra L, D'Alba G, Resta O (2016) An electronic nose may sniff out amyotrophic lateral sclerosis. Respir Physiol Neurobiol 232:22–25

Gardner JW, Bartlett PN (1999) Electronic noses—principles and applications. Oxford University Press, Oxford, UK

Gardner JW, Vincent TA (2016) Electronic noses for well-being: breath analysis and energy expenditure. Sensors 16:947

Guo W, Gan F, Kong H, Wu J (2015) Signal model of electronic noses with metal oxide semiconductor. Chemometr Intell Lab Syst 143:130–135

Kollár I (1994) Frequency domain system identification toolbox for use with MATLAB. The MathWorks Inc., Natick, MA

Qiu S, Wang J (2017) The prediction of food additives in the fruit juice based on electronic nose with chemometrics. Food Chem 230:208–214

Romero-Flores A, McConnell LL, Hapeman CJ, Ramirez M, Torrents A (2017) Evaluation of an electronic nose for odorant and process monitoring of alkaline-stabilized biosolids production. Chemosphere 186:151–159

Rosenqvist F, Tan AH, Godfrey KR, Karlström A (2006) Direction-dependent system modeling approaches exemplified through an electronic nose system. IEEE Trans Control Syst Technol 14:526–531

Tan AH (2009) Direction-dependent systems—a survey. Automatica 45:2729–2743

Tan AH, Godfrey KR (2001) Identification of processes with direction-dependent dynamics. IEE Proc Control Theory Appl 148:362–369

Tan AH, Godfrey KR (2004) Modeling of direction-dependent processes using Wiener models and neural networks with nonlinear output error structure. IEEE Trans Instrum Meas 53:744–753

Wei Z, Wang J, Zhang W (2015) Detecting internal quality of peanuts during storage using electronic nose responses combined with physicochemical methods. Food Chem 177:89–96

Wojnowski W, Majchrzak T, Dymerski T, Gębicki J, Namieśnik J (2017) Electronic noses: powerful tools in meat quality assessment. Meat Sci 131:119–131

Zhao H-T, Pang K-Y, Lin W-L, Wang Z-J, Gao D-Q, Guo M-J, Zhuang Y-P (2016) Optimization of the *n*-propanol concentration and feedback control strategy with electronic nose in erythromycin fermentation processes. Process Biochem 51:195–203

Chapter 7
Case Study on the Identification of a Multivariable Cooling System with Time-Varying Delay

7.1 Description of System and Experimental Setup

The problem of time-varying delay has received increasing attention in the control community (Wu et al. 2014; Ahn et al. 2015; Zeng et al. 2015). Such a delay significantly increases the complexity involved in identification and control. Examples of practical systems which exhibit such characteristics are solar air conditioning plants where the time delay is caused by fluid transport time (Garcia-Gabin et al. 2009), cutting machines where the rotational speed determines the delay (Michiels et al. 2005), networked control systems where the time delay is caused by the transmission of data through communication networks (Zhao et al. 2010), electronic neural networks where non-ideal components lead to time-varying delay (Chen et al. 2006), injection moulding processes where the process state delay varies within a bounded interval (Wang et al. 2013) and mist reactor systems where the time-varying delay could be caused by condensation (Cham et al. 2017).

The system under test in this Case Study is a thermodynamic cooling system described in Cham et al. (2010b), developed for applications in metallurgy and food industry. Further details of the experiment are reported in Tan and Cham (2011). The hardware comprises a flow control system, a Peltier system and a cool air inducer connected to a cooling chamber. In the system, ambient air at 1.8 bar is supplied using an air pump. A stepper motor controls the position of a valve which is used to adjust the flow rate. The amount of heating is controlled by the duty ratio of a PWM signal supplied to the Peltier module. These two inputs affect the temperature in the cooling chamber. This chamber is cylindrical in shape, having a height of 75 mm, an outer diameter of 100 mm and a wall thickness of 7 mm.

In the experiment, uncorrelated multisines with period $N = 600$ were applied. The multisines had uniform DFT magnitude at the excited harmonics as follows:

$$\text{Signal A: } \gamma_{\text{Signal\_A}} = \{1, 7, 13, \ldots, 235\}; \tag{7.1}$$

$$\text{Signal B: } \gamma_{\text{Signal\_B}} = \{5, 11, 17, \ldots, 239\}. \tag{7.2}$$

A. H. Tan and K. R. Godfrey, *Industrial Process Identification*, Advances in Industrial Control, https://doi.org/10.1007/978-3-030-03661-4_7

For the testing set, Signal A was applied as u_1 and Signal B was applied as u_2. For the validation set, Signal A was applied as u_2 and Signal B was applied as u_1.

The sampling frequency was limited by the presence of slew rate nonlinearity in the stepper motor. The maximum rate of change was 0.05 s per step of size 7.5° (Cham et al. 2010b). A sampling frequency of 1 Hz was selected in order not to excite the slew rate nonlinearity. This setting was made considering that there were 20 steps from the minimum level of 75° to the maximum level of 225° for u_1, the range itself also being selected to coincide with the linear range of the system. The amplitude range for u_2 was from 0 to 100% duty ratio. Four steady-state periods of the output were collected for both testing and validation sets.

7.2 Identification of Linear Model

The contributions of each input to the output were extracted by considering only the harmonics at which each input was excited. The FRFs defined by $G_i(z^{-1}) = Y(z^{-1})/U_i(z^{-1})$ with $i = 1, 2$, are plotted in Figs. 7.1 and 7.2 for the four individual steady-state periods and the averaged period of the testing set. From Fig. 7.1, an interesting observation is seen in the magnitude of the FRF, where the averaged FRF shows a different trend compared with those of the individual periods. This is due to the variation in the FRF phase between the individual periods where the phase shifts become increasingly negative going from the first period down to the fourth period. This leads to a cancellation of the contributions from the individual periods resulting in a dip in the magnitude of the FRF at harmonic 181. Such a trend is not seen in Fig. 7.2 where the averaged period is a good representation of the FRF of the individual periods.

The power at both excited and non-excited harmonics was checked by plotting the 2400-point DFT following Step 3 in the procedure in Sect. 5.4.1. The plot is shown in Fig. 7.3. In particular, contributions from the linear and nonlinear terms are limited to harmonics which are integer multiples of $P = 4$, since the DFT was taken across four periods. The linear contributions corresponding to inputs u_1 and u_2 appear at harmonics $\{1, 7, 13, \ldots, 235\} \times 4 = \{4, 28, 52, \ldots, 940\}$ and $\{5, 11, 17, \ldots, 239\} \times 4 = \{20, 44, 68, \ldots, 956\}$, respectively. The odd order nonlinear distortion appears at $P \times$ (non-excited harmonics) $= \{3, 9, 15, \ldots, 297\} \times 4 = \{12, 36, 60, \ldots, 1188\}$, listing only the harmonics up to the Nyquist frequency. The even order nonlinear distortion appears at $\{2, 4, 6, \ldots, 300\} \times 4 = \{8, 16, 24, \ldots, 1200\}$.

Disturbances such as noise and time-varying contributions appear at all the harmonics. These disturbances were contributed by fluctuations in the ambient temperature as well as turbulent flow in the cool air inducer. The latter was caused, or perhaps aggravated, by the rough surface of the air tunnel. From Fig. 7.3, both odd order nonlinear distortion (represented by magenta squares) and even order nonlinear distortion (represented by blue plusses) can be considered negligible as they generally do not rise above the disturbance floor. In contrast, the effects of disturbance (represented by red crosses) are rather significant.

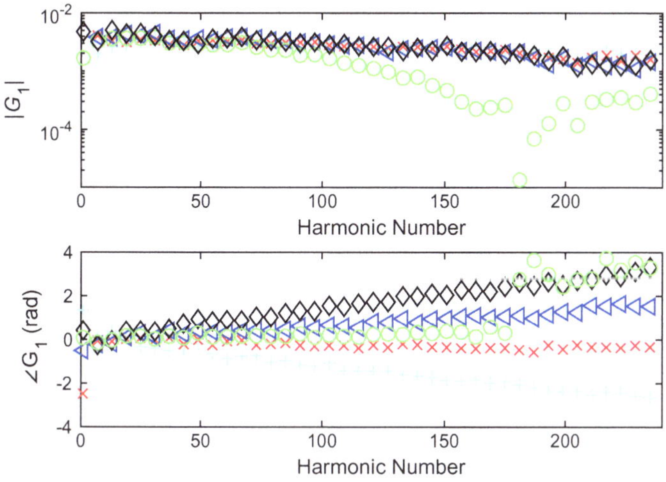

Fig. 7.1 Measured FRF of the system due to input to the flow control system. Black diamonds: first period; blue triangles: second period; red crosses: third period; cyan plusses: fourth period; green circles: average of four individual steady-state periods

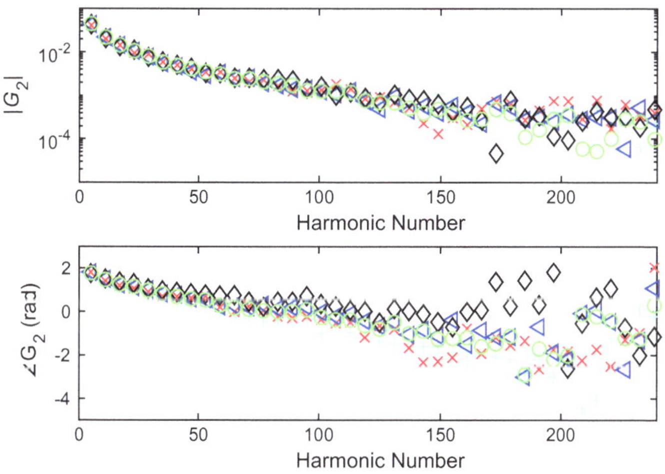

Fig. 7.2 Measured FRF of the system due to input to the Peltier system. Black diamonds: first period; blue triangles: second period; red crosses: third period; cyan plusses: fourth period; green circles: average of four individual steady-state periods

Linear models were then identified using maximum likelihood estimation in the frequency domain. This led to the transfer functions (Cham et al. 2010a)

$$G_1(z^{-1}) = \frac{2.476 \times 10^{-3}}{1 - 0.1597z^{-1}}, \tag{7.3}$$

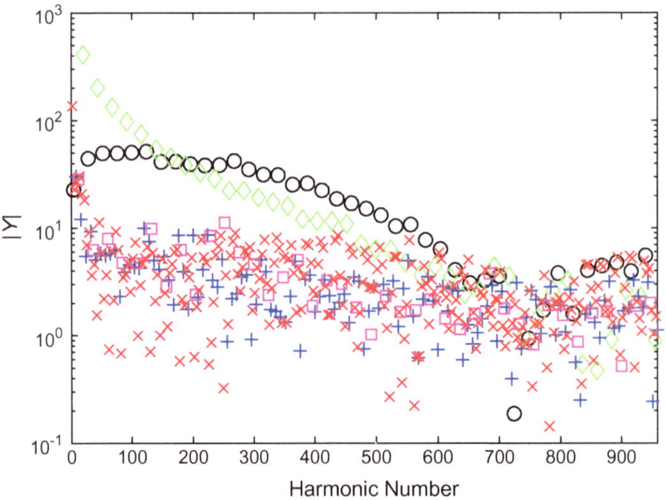

Fig. 7.3 Measured 2400-point DFT. Black circles: power at harmonics {4, 28, 52, ..., 940}; green diamonds: power at harmonics {20, 44, 68, ..., 956}; magenta squares: power at harmonics {12, 36, 60, ...}; blue plusses: power at harmonics {8, 16, 24, ...}; red crosses: power at the some of the remaining harmonics (not all are shown in order to improve clarity)

$$G_2(z^{-1}) = \frac{3.924 \times 10^{-5} - 4.637 \times 10^{-4} z^{-1} - 5.811 \times 10^{-4} z^{-2}}{1 - 1.752 z^{-1} + 0.930 z^{-2} - 0.171 z^{-3}}. \tag{7.4}$$

For G_1, this was obtained by fitting only a single period of data as several complications were encountered in fitting the averaged data due to the effects of time variation. In particular, when the averaged data were utilised, the model estimated had a very high order of 33 and was badly conditioned, with the model gain increasing rapidly near the Nyquist frequency.

The performance of the linear models is shown in Figs. 7.4 and 7.5. From Fig. 7.4, for the channel involving the flow control system, the linear model is very poor and the errors obtained are very large compared with the actual output. The model output is much more oscillatory compared to the actual one. The inability of the linear model to describe G_1 is to be expected, since the varying FRF phase indicates that there is likely to be a significant time-varying component in the form of time-varying delay. The presence of the time-varying component was also confirmed using the frequency domain indicator for determining the main source of disturbance in the system (Lataire et al. 2012) [see Eqs. 5.10 and 5.11 as well as further details given in Tan et al. (2015)]. In contrast, the linear model for the channel involving the Peltier system G_2 performs very well. The match in Fig. 7.5 is excellent with the error being much smaller than the actual output.

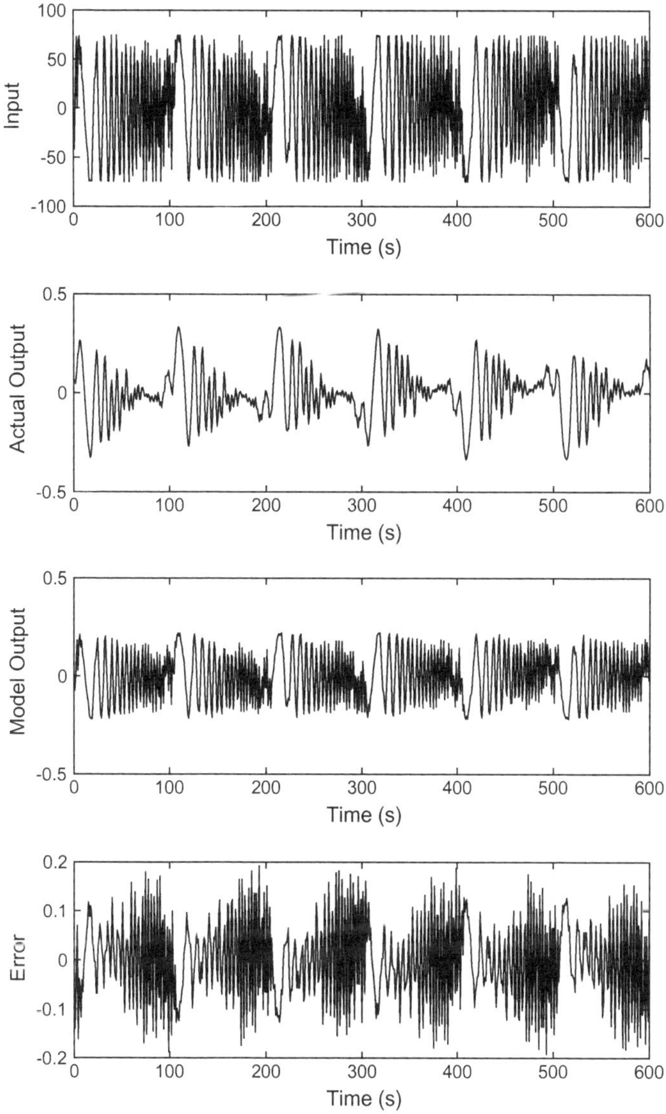

Fig. 7.4 Input, actual output, model output and error signals for G_1

7.3 Delay Reconciliation Using Crosscorrelation

From Fig. 7.1 as well as the results of Sect. 7.2, the averaged magnitude of the FRF due to the flow control system is not useful for estimating the linear dynamics since the shape is not an accurate representation of the magnitude response of the actual system. The cause of the problem can be guessed from the difference in the

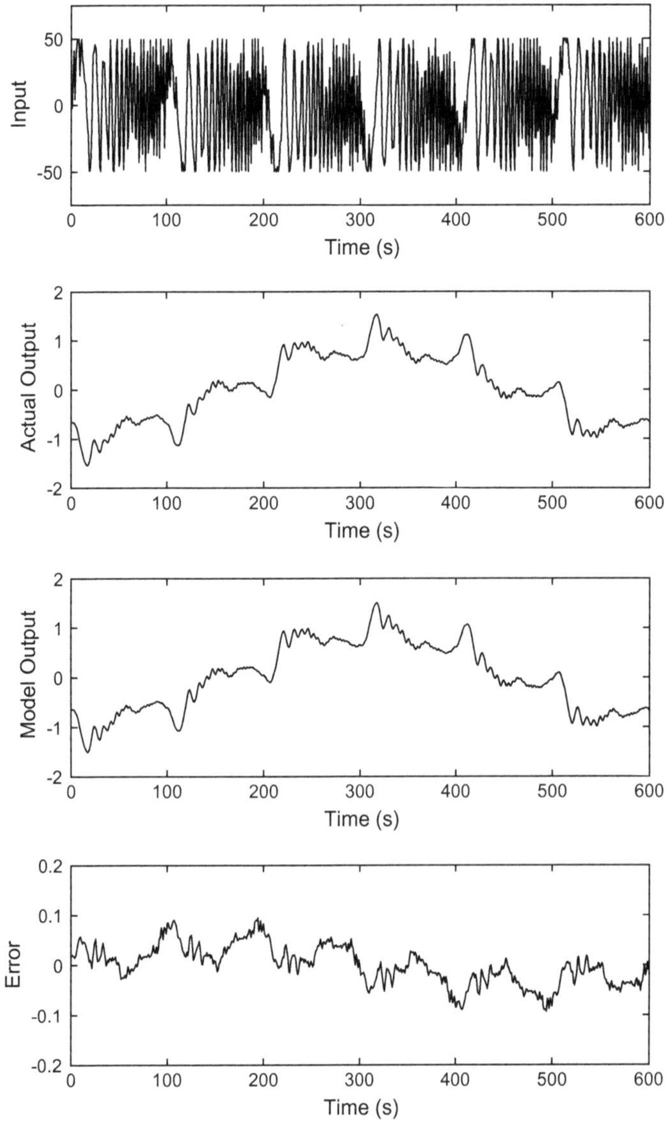

Fig. 7.5 Input, actual output, model output and error signals for G_2

FRF phase of Fig. 7.1; it is likely to be attributed to time-varying delay. Hence, a technique called delay reconciliation (Tan and Cham 2011) was applied to remove the relative delays between the individual periods, 'reconciling' them, so that the averaged period becomes more representative of the individual periods. The discrete values of the delays were detected using crosscorrelation. (Another recent work making use of correlation in the estimation of delay is described by Li et al. (2017).)

Even though the reconciliation in the time domain is applicable only to discrete values of delay and the effectiveness of the method depends on the actual delay in the system, the technique is simple and provides a good starting point for subsequent analysis. Further compensation and modelling of the remaining delay can be achieved in a later step.

The steps in delay reconciliation (Tan and Cham 2011) are summarised as follows:

Step 1 The 'optimal' delay in period m, denoted by d_m, relative to a reference period m_{ref} is calculated based on

$$d_m = \arg\max_{\tau} \left[\sum_{i=1}^{N} y_{m_{\text{ref}}}(i) y_m(\text{mod}(i + \tau - 1, N) + 1) \right];$$
$$i = 1, 2, \ldots, N; \tau \in Z; \forall m : \{1 \le m \le m_{\text{total}}, m \in Z\} \tag{7.5}$$

where y_m is the output of period m, $y_{m_{\text{ref}}}$ is the output of the reference period, m_{total} is the number of periods measured, N is the signal period, Z denotes the set consisting of integers and mod(v, w) represents the modulo operator giving the remainder after the division of v by w. For $m = m_{\text{ref}}$, $d_m = 0$ by definition.

Step 2 To shift y_m by its optimal delay, set

$$y_m(i) = y_m(\text{mod}(i + d_m - 1, N) + 1) \tag{7.6}$$

$y_{m_{\text{ref}}}$ is left unchanged by this operation.

Equation 7.5 essentially finds the delay at which the crosscorrelation function is maximum. Choosing $m_{\text{ref}} = 1$, the crosscorrelation functions are plotted in Fig. 7.6, noting that $N = 600$ in this example. The maximum value of the crosscorrelation between the first period and the second period occurs at $\tau = 1$ thus giving $d_2 = 1$ s. For the crosscorrelation between the first period and the third period, the peak occurs at $\tau = 2$ which gives $d_3 = 2$ s. However, the crosscorrelation is also large at $\tau = 1$, being only slightly smaller than that at $\tau = 2$; this points to the need to consider fractional delays later. Similarly, the crosscorrelation between the first period and the fourth period reaches maximum at $\tau = 2$ which gives $d_4 = 2$ s. In this case, the value at $\tau = 3$ is only slightly smaller.

The codes for obtaining Fig. 7.6 using MATLAB are given below. (MATLAB® is a registered product of The MathWorks, Inc.)

```
%define y1, y2, y3 and y4 as the outputs corresponding
%to the first, second, third and fourth individual
%periods
%define signal period
N=600;

%compute crosscorrelation between 1 st and 2nd periods
R2=real(ifft(conj(fft(y1)).*fft(y2)))/N;
plot([-20:20],[R2(581:600);R2(1:21)],'k<')
```

```
hold
plot([-20:20],[R2(581:600);R2(1:21)],'k')
%compute crosscorrelation between 1st and 3rd periods
R3=real(ifft(conj(fft(y1)).*fft(y3)))/N;
plot([-20:20],[R3(581:600);R3(1:21)],'kx')
plot([-20:20],[R3(581:600);R3(1:21)],'k')
%compute crosscorrelation between 1st and 4th periods
R4=real(ifft(conj(fft(y1)).*fft(y4)))/N;
plot([-20:20],[R4(581:600);R4(1:21)],'ks')
plot([-20:20],[R4(581:600);R4(1:21)],'k')
xlabel('Delay')
ylabel('Crosscorrelation Function')
```

The data for the second, third and fourth periods were therefore adjusted according to Eq. 7.6 by removing the relative delays. These periods were advanced by 1 s, 2 s and 2 s, respectively. It is worth highlighting that the collected data also initially exhibited some non-causal dynamics due to a feedforward mechanism in the flow control system. This can be observed from the positive phase shifts in the FRF phase plot in Fig. 7.1. The non-causality was removed, in addition to (or along with) the delay reconciliation step, before further analysis was carried out.

The FRF after delay reconciliation is plotted in Fig. 7.7, where it can be seen that now the averaged data are more representative of the individual periods of data. The large variations between the individual periods in the phase plot in Fig. 7.1 have been significantly reduced in Fig. 7.7; however, they are not completely eliminated. The positive phase shift caused by the non-causality has also been removed.

Delay reconciliation therefore allows partial removal of the effects of the time-varying delay so that the identification of the time-invariant part of the channel becomes an easier task (see Sect. 7.4). The averaged signal after delay reconciliation is given by

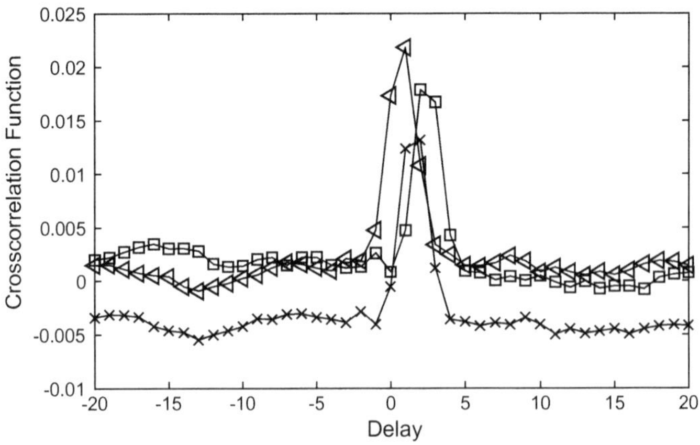

Fig. 7.6 Crosscorrelation functions of the first period with the second period (triangles), the third period (crosses) and the fourth period (squares) of the testing data (Reproduced with permission from Tan and Cham © 2011 IET)

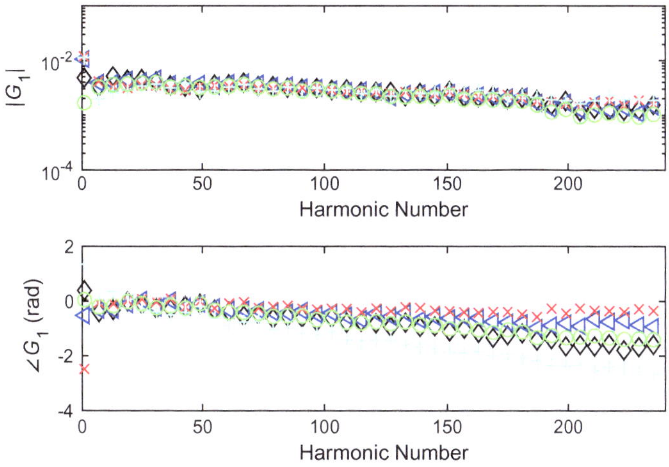

Fig. 7.7 Measured FRF of the system due to input to the flow control system after delay reconciliation and removal of non-causality. Black diamonds: first period; blue triangles: second period; red crosses: third period; cyan plusses: fourth period; green circles: average of four individual steady-state periods (Reproduced with permission from Tan and Cham © 2011 IET)

Table 7.1 Statistics of the system disturbance before and after delay reconciliation

Period	Mean squared value before delay reconciliation	Mean absolute value before delay reconciliation	Mean squared value after delay reconciliation	Mean absolute value after delay reconciliation
First	1.22×10^{-2}	9.08×10^{-2}	1.28×10^{-3}	2.98×10^{-2}
Second	1.13×10^{-2}	8.71×10^{-2}	6.66×10^{-3}	7.05×10^{-2}
Third	1.79×10^{-2}	1.10×10^{-1}	1.47×10^{-2}	1.03×10^{-1}
Fourth	2.20×10^{-2}	1.12×10^{-1}	1.42×10^{-2}	9.73×10^{-2}
Overall	1.59×10^{-2}	1.02×10^{-1}	9.20×10^{-3}	7.53×10^{-2}

$$y_a(i) = \frac{1}{P} \sum_{q=0}^{P-1} y(\mathrm{mod}(i-1, N) + 1 + Nq); \quad i = 1, 2, \ldots, N \tag{7.7}$$

where P is the number of steady-state periods measured ($P = 4$ in this example).

It is of interest to analyse the disturbance signals both before and after delay reconciliation. The disturbance signal is defined by

$$n(i) = y(i) - y_a(\mathrm{mod}(i-1, N) + 1); \quad i = 1, 2, \ldots, NP \tag{7.8}$$

The signals for the testing set are plotted in Fig. 7.8. The statistics in terms of the mean squared value and the mean absolute value are summarised in Table 7.1.

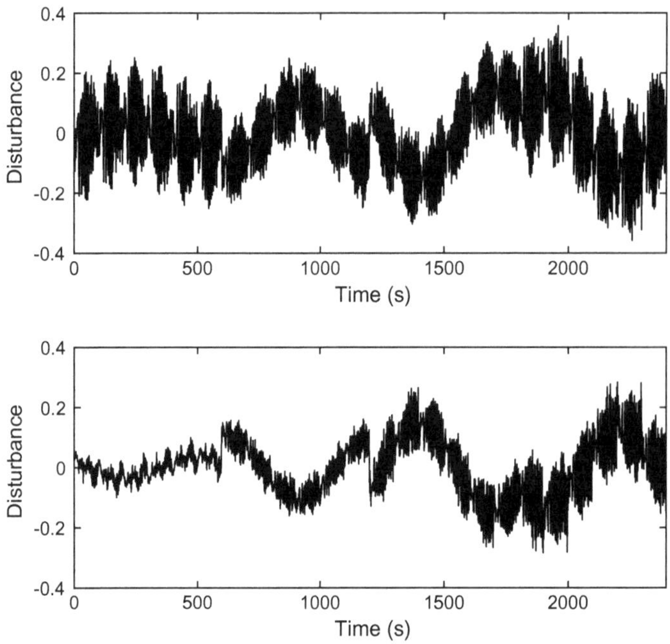

Fig. 7.8 Disturbance signal before delay reconciliation (top) and after delay reconciliation (bottom)

From Fig. 7.8 as well as Table 7.1, the amount of disturbance has decreased after delay reconciliation. The amount of disturbance is the least in the first period after delay reconciliation, as this period was chosen as the reference period. The disturbance is still significant in the third and fourth periods, mainly because the delay reconciliation technique in the time domain deals only with integer values of delay. Indeed, if the delay reconciliation was performed in the frequency domain, fractional shifts can be accommodated. However, that will still necessitate the treatment of Sect. 7.5 for dealing with varying values of delay within individual periods and hence, the delay reconciliation was performed only for integer delays.

Fractional relative delay values can be identified in the frequency domain if necessary. For example, the relationship for the relative delay between the first and second periods is $Y_{\text{2nd\_period}}(\mathrm{j}\omega) = \mathrm{e}^{-\mathrm{j}\omega\tau} Y_{\text{1st\_period}}(\mathrm{j}\omega)$. The corresponding phase equation is $\omega\tau = \angle Y_{\text{1st\_period}}(\mathrm{j}\omega) - \angle Y_{\text{2nd\_period}}(\mathrm{j}\omega)$. The 'best' value of delay τ can be solved using least squares. In MATLAB, the following codes can be applied.

```
%define excited harmonics and excited frequencies
f=[1:6:235];
w=f'/600*2*pi;
%obtain the DFTs of the first and second periods
Y1=fft(y1);Y2=fft(y2);
difference=angle(Y1(f+1))-angle(Y2(f+1));
%unwrap to obtain smooth curve
```

```
difference_unwrapped=unwrap(difference);
delay=w\difference_unwrapped;
```

Using the codes above (with minor adjustments for the third and fourth periods) resulted in the relative delays of 0.696, 1.502 and 2.420 for the second, third and fourth periods, respectively.

7.4 Offline Identification of Invariant Dynamics

Maximum likelihood estimation was applied to the averaged period in order to estimate the invariant dynamics as detailed in Tan and Cham (2011). The MSE and mean absolute error (MAE) defined by

$$\mathrm{MSE} = \frac{1}{2400}\sum_{i=1}^{2400}\left(\hat{y}(i) - y(i)\right)^2, \tag{7.9}$$

$$\mathrm{MAE} = \frac{1}{2400}\sum_{i=1}^{2400}\left|\hat{y}(i) - y(i)\right| \tag{7.10}$$

were computed, where $\hat{y}$ denotes the model output. The model order was selected to give the minimum MSE and MAE on the validation set. The delay was fixed at zero as this achieved lower error values compared to when the delay was also left as a parameter to be estimated.

Two methods were applied to estimate the continuous-time model; such a model was sought after as it is more amenable to adaptive identification of delay which will be performed later. The first method uses ELiS available in the Frequency Domain System Identification Toolbox (Kollár 1994) to directly identify a continuous-time model. This resulted in

$$G_{\text{direct}}(s) = \frac{5.87\times 10^{-3}s^3 - 2.49\times 10^{-3}s^2 + 5.19\times 10^{-2}s + 7.48\times 10^{-4}}{s^3 + 8.60s^2 + 14.9s + 3.74\times 10^{-1}}. \tag{7.11}$$

The second method estimates a discrete-time model using ELiS and then converts it to a continuous-time model via ZOH transformation. This indirect method led to

$$G_{\text{indirect}}(z^{-1}) = \frac{1.46\times 10^{-3} + 9.24\times 10^{-5}z^{-1} - 1.51\times 10^{-3}z^{-2}}{1 - 1.12z^{-1} + 1.40\times 10^{-1}z^{-2}}, \tag{7.12}$$

$$G_{\text{indirect}}(s) = \frac{1.46\times 10^{-3}s^2 + 6.83\times 10^{-3}s + 1.03\times 10^{-4}}{s^2 + 1.97s + 5.23\times 10^{-2}}. \tag{7.13}$$

It is interesting to note that $G_{\text{indirect}}(s)$ is of a lower order than $G_{\text{direct}}(s)$. This is because setting the model order of $G_{\text{indirect}}(z^{-1})$ to three zeros and three poles resulted in an unstable transfer function.

The fractional delay that remains causes a lowpass filtering effect. The true Fourier transform $Y_{\text{true}}(\omega)$ is scaled by a frequency-dependent factor to give $Y_a(\omega)$. This scaling is described by

$$Y_a(\omega) = Y_{\text{true}}(\omega)\left[1 + \mathrm{j}\omega\mu - \frac{\omega^2(\sigma^2 + \mu^2)}{2} + \cdots\right] \tag{7.14}$$

where μ and σ^2 are the mean and variance, respectively, of the remnant uncompensated delay (Souders et al. 1990). This means that $G_{\text{direct}}(s)$, $G_{\text{indirect}}(z^{-1})$ and $G_{\text{indirect}}(s)$ are in fact biased estimates. Thus, the identification was repeated taking into account of the scaling to give unbiased estimates. The values of $\mu = 0$ and $\sigma^2 = 1/12$ were applied, assuming uniform error distribution of the delay. This led to (Tan and Cham 2011)

$$G_{\text{direct\_unbiased}}(s) = \frac{3.39 \times 10^{-3}s^3 - 1.34 \times 10^{-3}s^2 + 3.70 \times 10^{-2}s + 5.29 \times 10^{-4}}{s^3 + 6.13s^2 + 10.6s + 2.64 \times 10^{-1}}, \tag{7.15}$$

$$G_{\text{indirect\_unbiased}}(z^{-1}) = \frac{1.49 \times 10^{-3} + 2.88 \times 10^{-4}z^{-1} - 1.73 \times 10^{-3}z^{-2}}{1 - 1.05z^{-1} + 7.74 \times 10^{-2}z^{-2}}, \tag{7.16}$$

$$G_{\text{indirect\_unbiased}}(s) = \frac{1.49 \times 10^{-3}s^2 + 9.00 \times 10^{-3}s + 1.30 \times 10^{-4}}{s^2 + 2.56s + 6.49 \times 10^{-2}}. \tag{7.17}$$

The estimated FRFs for $G_{\text{direct}}(s)$ and $G_{\text{indirect}}(s)$ are shown in Tan and Cham (2011). Those for $G_{\text{direct\_unbiased}}(s)$ and $G_{\text{indirect\_unbiased}}(s)$ are in fact very similar to their biased counterparts. The FRFs are close to one another below 0.5 rad s^{-1}. However, the deviation is significant at high frequencies. For example, a much larger negative phase response is obtained for the direct estimation method.

The zeros and poles of the estimated FRFs are depicted in Fig. 7.9. For all four FRFs, there is a zero-pole pair close to the origin which nearly cancels out (but they do not cancel out, only nearly). $G_{\text{direct}}(s)$ and $G_{\text{direct\_unbiased}}(s)$ have zeros on the right hand side of the s-plane, which corresponds to non-minimum phase behaviour. Interestingly, it was highlighted by Chen and Peng (2005) that the thermodynamic system which they researched on, in the form of a fuel cell membrane humidifier, also exhibited such behaviour. According to Chen and Peng (2005), this phenomenon can be attributed to the interrelation between vapour mass flow, humidity and temperature. Additionally, in going from $G_{\text{direct}}(s)$ to $G_{\text{direct\_unbiased}}(s)$, two of the real poles have become a pair of complex conjugate poles.

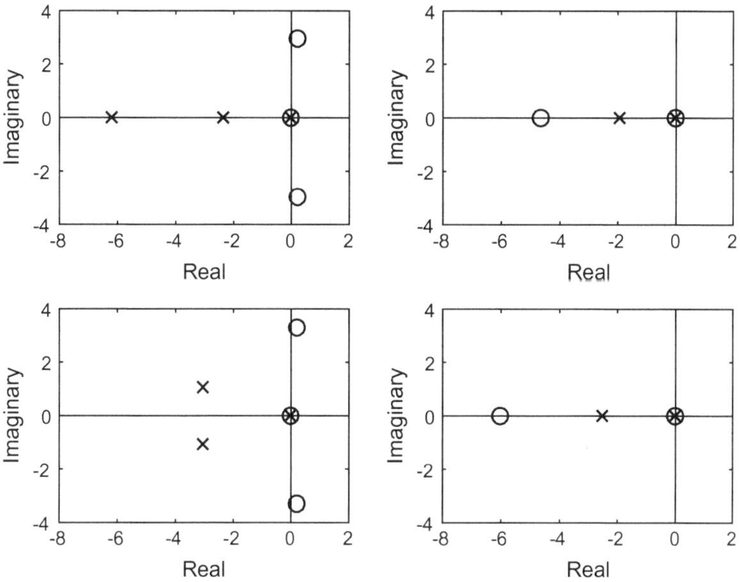

Fig. 7.9 Zeros and poles of the estimated FRFs. Top left: $G_{\mathrm{direct}}(s)$; top right: $G_{\mathrm{indirect}}(s)$; bottom left: $G_{\mathrm{direct\_unbiased}}(s)$; bottom right: $G_{\mathrm{indirect\_unbiased}}(s)$

The step and impulse responses for $G_{\mathrm{direct\_unbiased}}(s)$ and $G_{\mathrm{indirect\_unbiased}}(s)$ are plotted in Fig. 7.10. While both the step responses are quite similar to each other, the impulse responses are significantly different. The reason behind this observation can be explained as follows. Both systems have a large magnitude fast component and a small magnitude slow component. The former is quite different in the two models but the latter is close to each other. This is consistent with the large deviation between the models in the frequency domain at high frequencies. Part of the problem was caused by the sampling frequency being limited by the slew rate nonlinearity. In addition to this, $G_{\mathrm{direct\_unbiased}}(s)$ is non-minimum phase but $G_{\mathrm{indirect\_unbiased}}(s)$ is not.

7.5 Online Adaptive Identification of Variable Delay

The adaptive identification was carried out using the gridding approach (Tan and Cham 2011) where the delay values were tested on a grid from 0 and 3 s. This range was selected based on a priori information. A grid size of 0.1 s was chosen as this was deemed a good trade-off between accuracy and computational load.

The adaptive algorithm works as follows:

Step 1 Break the time domain data into subrecords of length L. There are thus $N_s = 2400/L$ subrecords in total.

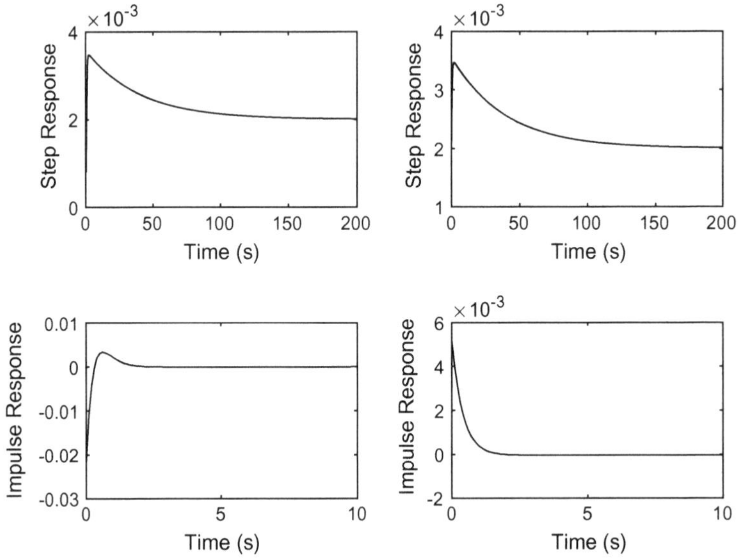

Fig. 7.10 Step response and impulse response of the estimated FRFs. Left: $G_{\text{direct\_unbiased}}(s)$; right: $G_{\text{indirect\_unbiased}}(s)$

Step 2 Initialise the delay in the first subrecord to 0 such that $D_1 = 0$.

Step 3 Set $m = 1$ and compute the error function

$$e(\theta_m) = \sum_{i=1}^{L} \left(\hat{y}(L(m-1)+i, \theta_m) - y(L(m-1)+i)\right)^2 \tag{7.18}$$

for all values of θ_m such that $0 \leq \theta_m \leq 3,\ \theta_m = 0.1p,\ p \in \mathrm{Z}$. In Eq. 7.18, $\hat{y}$ is the model output (using $G_{\text{direct\_unbiased}}(s)$ or $G_{\text{indirect\_unbiased}}(s)$ since the effect of fractional delay is assumed to be compensated by the adaptive technique).

Step 4 Set D_{m+1} according to

$$D_{m+1} = \arg\min_{\theta_m}[e(\theta_m)]. \tag{7.19}$$

Step 5 Increment m by 1 and repeat Steps 3–5 until $m = N_s - 1$.

Results for varying the subrecord length L from 10 to 60 are plotted in Fig. 7.11. Within this range, for $G_{\text{direct\_unbiased}}(s)$, the smallest MSE and MAE values are achieved with $L = 50$ whereas for $G_{\text{indirect\_unbiased}}(s)$, the smallest MSE and MAE values are achieved with $L = 40$. In general, increasing L will make the adaptive response slower but decreasing L will increase its sensitivity to other forms of disturbance.

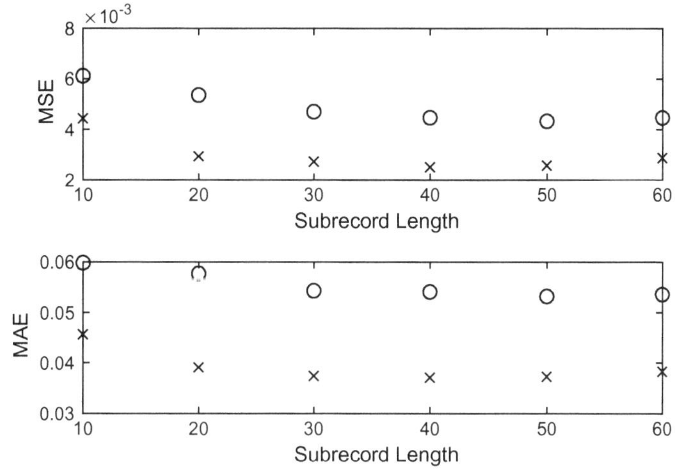

Fig. 7.11 The effect of varying subrecord length on the validation data. Circles: $G_{\text{direct\_unbiased}}(s)$; crosses: $G_{\text{indirect\_unbiased}}(s)$

Table 7.2 Best-case MSE and MAE values achieved using several continuous-time models on the validation set

Method	MSE ($\times 10^{-3}$)	MAE ($\times 10^{-2}$)
Direct, non-adaptive	5.58	5.96
Indirect, non-adaptive	5.68	6.18
Direct, adaptive	4.33	5.32
Indirect, adaptive	2.51	3.71

The best-case MSE and MAE values for both direct and indirect methods with and without the adaptive approach are summarised in Table 7.2. While the direct method outperformed the indirect one in the absence of adaptation, the opposite is true when adaptive algorithm was applied.

To understand why the indirect method finally attained a better performance compared to the direct one, the estimated FRFs are plotted in Fig. 7.12 for $G_{\text{direct\_unbiased}}(s)$ and $G_{\text{indirect\_unbiased}}(s)$ superimposed on the validation data. While both fit the magnitude data quite well with $G_{\text{indirect\_unbiased}}(s)$ giving a slightly better match, $G_{\text{direct\_unbiased}}(s)$ has a closer resemblance to the phase data. This is probably caused by the higher order model identified using the direct technique. In the absence of adaptive delay, $G_{\text{direct\_unbiased}}(s)$ therefore achieved smaller error values. However, when the delay was estimated online, $G_{\text{indirect\_unbiased}}(s)$ had a larger scope for improvement as the delay directly affected the phase shift. This compensated for the poorer fit of $G_{\text{indirect\_unbiased}}(s)$ to the phase data which finally led to the superior performance of the indirect technique over its direct counterpart.

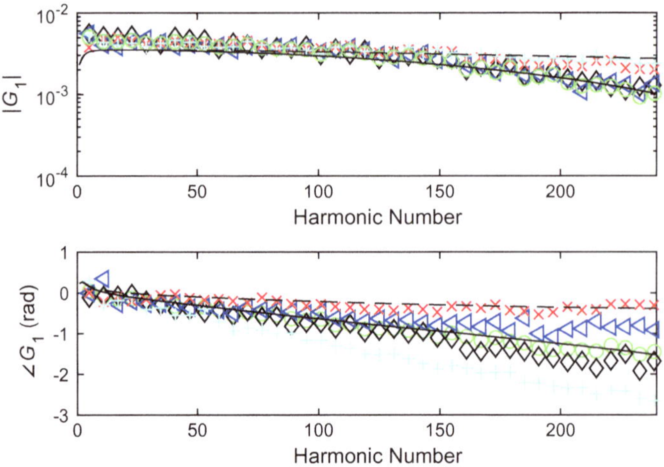

Fig. 7.12 Estimated FRFs superimposed on validation data. Solid line: $G_{\text{direct\_unbiased}}(s)$; dashed line: $G_{\text{indirect\_unbiased}}(s)$. Black diamonds: first period; blue triangles: second period; red crosses: third period; cyan plusses: fourth period; green circles: average of four individual steady-state periods (Reproduced with permission from Tan and Cham © 2011 IET)

The experiment and model outputs for $G_{\text{indirect\_unbiased}}(s)$, with $L = 40$, are shown in Fig. 7.13. This corresponds to the best model obtained for the channel connecting the flow control system to the cooling chamber. In comparison with Fig. 7.4, it can be seen that the error has reduced considerably with the help of delay reconciliation as well as online adaptive identification of the variable delay.

This Case Study has illustrated that the application of a carefully designed set of perturbation signals can successfully lead to the identification of models which are able to capture the important characteristics of the system under test. The use of periodic signals and the collection of several periods of the steady-state output enable detailed analysis of important characteristics of the system. Both continuous-time and discrete-time models can be estimated, depending on the required application.

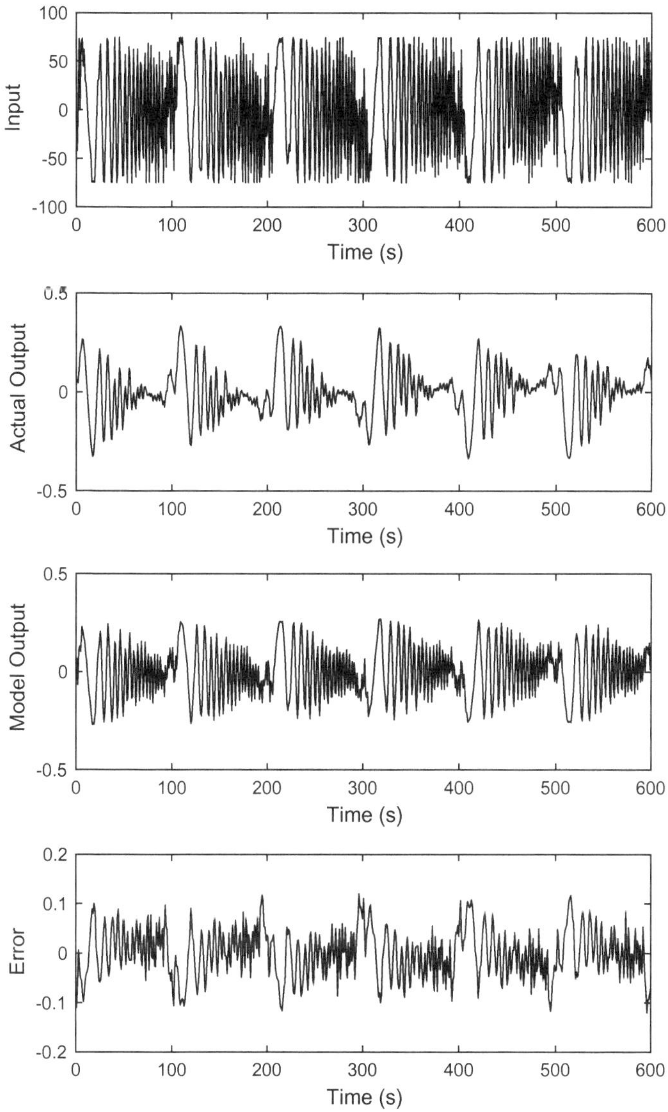

Fig. 7.13 Input, actual output, model output and error signals for G_1 applying online adaptive identification

References

Ahn CK, Shi P, Wu L (2015) Receding horizon stabilization and disturbance attenuation for neural networks with time-varying delay. IEEE Trans Cybern 45:2680–2692

Cham CL, Tan AH, Ramar K (2010a) Modelling of a hyperfast switching Peltier cooling system using a variable time delay model. In: Proceedings of the UKACC international conference on control "Control 2010", Coventry, UK, 7–10 Sept, pp 179–184

Cham CL, Tan AH, Tan WH (2010b) Hyperfast switching Peltier cooling system benchmark. In: Proceedings of the UKACC international conference on control "Control 2010", Coventry, UK, 7–10 Sept, pp 185–190

Cham CL, Tan AH, Tan WH (2017) Identification of a multivariable nonlinear and time-varying mist reactor system. Control Eng Pract 63:13–23

Chen D, Peng H (2005) Analysis of non-minimum phase behavior of PEM fuel cell membrane humidification systems. In: Proceedings of the American control conference (Paper FrA14.2), Portland, OR, 8–10 June

Chen W-H, Lu X, Guan Z-H, Zheng WX (2006) Delay-dependent exponential stability of neural networks with variable delay: an LMI approach. IEEE Trans Circ Syst—II: Exp Briefs 53:837–842

Garcia-Gabin W, Zambrano D, Camacho EF (2009) Sliding mode predictive control of a solar air conditioning plant. Control Eng Pract 17:652–663

Kollár I (1994) Frequency domain system identification toolbox for use with MATLAB. The MathWorks Inc., Natick

Lataire J, Louarroudi E, Pintelon R (2012) Detecting a time-varying behavior in frequency response function measurements. IEEE Trans Instrum Meas 61:2132–2143

Li L-J, Dong T-T, Zhang S, Zhang X-X, Yang S-P (2017) Time-delay identification in dynamic processes with disturbance via correlation analysis. Control Eng Pract 62:92–101

Michiels W, Van Assche V, Niculescu S-I (2005) Stabilization of time-delay systems with a controlled time-varying delay and applications. IEEE Trans Autom Control 50:493–504

Souders TM, Flach DR, Hagwood C, Yang GL (1990) The effects of timing jitter in sampling systems. IEEE Trans Instrum Meas 39:80–85

Tan AH, Cham CL (2011) Continuous-time model identification of a cooling system with variable delay. IET Control Theory Appl 5:913–922

Tan AH, Cham CL, Godfrey KR (2015) Comparison of three modeling approaches for a thermodynamic cooling system with time-varying delay. IEEE Trans Instrum Meas 64:3116–3123

Wang L, Mo S, Qu H, Zhou D, Gao F (2013) H_∞ design of 2D controller for batch processes with uncertainties and interval time-varying delays. Control Eng Pract 21:1321–1333

Wu L, Yang X, Lam H-K (2014) Dissipativity analysis and synthesis for discrete-time T-S fuzzy stochastic systems with time-varying delay. IEEE Trans Fuzzy Syst 22:380–394

Zeng H-B, He Y, Wu M, She J (2015) Free-matrix-based integral inequality for stability analysis of systems with time-varying delay. IEEE Trans Autom Control 60:2768–2772

Zhao Y-B, Liu G-P, Rees D (2010) Stability and stabilisation of discrete-time networked control systems: a new time delay system approach. IET Control Theory Appl 4:1859–1866

Chapter 8
Software for Signal Design

8.1 *prs*

The *prs* routine can be freely downloaded from https://sites.google.com/view/signaldesign/prs. The routine runs on MATLAB®. (MATLAB® is a registered product of The MathWorks, Inc.) It incorporates functions to generate PRB signals from the classes of MLB, QRB, HAB and TPB, as well as near-binary signals from the class of QRT. It supports signal generation with period N up to 50,000. The signals may have all harmonics present or only odd harmonics present; in the latter case, the signals are inverse-repeat signals from the above classes. Functions are also available to calculate three measures of signal quality, namely, PIPS, PIPSE and EMINE. The routine is based on the work in Tan and Godfrey (2002). A subsidiary program allows the generation of direct synthesis ternary signals. These signals have harmonic multiples of two and three suppressed.

To run the program, type `prs` in the MATLAB command window. This will bring up a graphical user interface (GUI). Enter the desired harmonic specification, signal length and number of signal levels. The class of pseudorandom signals available will be displayed. Click on the desired class and then click 'Design'. The signal name can be changed if desired. An example for generating a QRB signal with $N = 59$ is shown in Fig. 8.1.

For MLB signals, the *prs* routine allows the user to specify a primitive polynomial. If the user chooses this option by clicking 'Options', another pop-up window will appear. For example, an MLB signal with $N = 63$ and characteristic equation $D^6 \oplus_2 D^5 \oplus_2 D^4 \oplus_2 D \oplus_2 1 = 0$ should be entered as '`1 1 0 0 1 1 1`' into the prompt for characteristic polynomial. The entry is to be interpreted as follows: D^0 term is present (1), D^1 term is present (1), D^2 term is absent (0), D^3 term is absent (0), D^4 term is present (1), D^5 term is present (1) and D^6 term is present (1). If the user chooses not to specify the primitive polynomial, then a default signal will be generated.

For MLB signals, the *prs* routine can determine the shifts due to the second-order shift-and-multiply property (Eq. 2.6) as well as the third-order shift-and-multiply

A. H. Tan and K. R. Godfrey, *Industrial Process Identification*, Advances in Industrial Control, https://doi.org/10.1007/978-3-030-03661-4_8

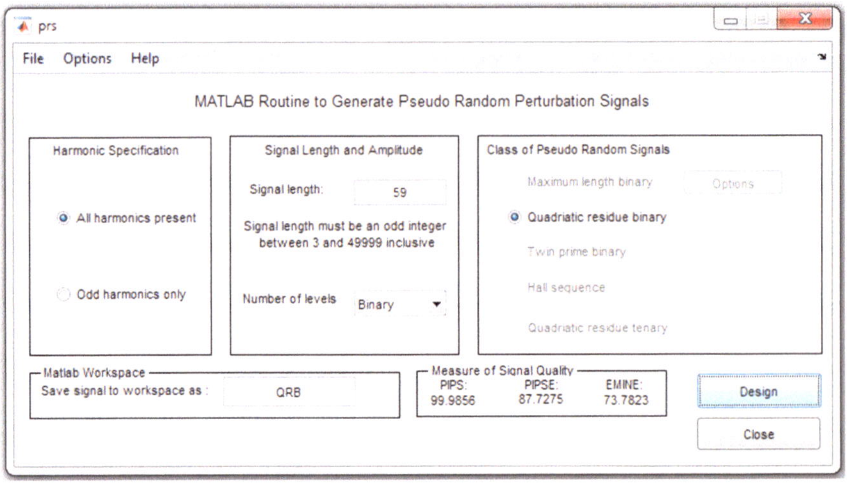

Fig. 8.1 Example of generating a QRB signal with $N = 59$ using *prs*

property (Eq. 2.9). (For inverse-repeat MLB signals, only the third-order shift-and-multiply property is applicable.) This function is available by clicking 'Options' which will bring up instructions in the MATLAB command window prompting input from the user through the keyboard. For example, for the MLB signal with characteristic equation $D^6 \oplus_2 D^5 \oplus_2 D^4 \oplus_2 D \oplus_2 1 = 0$, and with appropriate inputs from the user, the MATLAB command window will appear as shown below. (Inputs from the user are underlined.) Compare this with the results of the second- and third-order shifts in Tables 6.1 and 6.2.

```
Calculating second order shift:
Please enter an integer between 1 and 62 or enter -1 to
quit.
New Delay = 1
The resulting shift is 39.
New Delay = 2
The resulting shift is 15.

New Delay = 3
The resulting shift is 11.

New Delay = -1
```

```
Calculating third order shift:
Please enter two integers between 1 and 62 or enter -1
to quit.
The integers must be increasing in value and entered in
the format [integer1 integer2].
New Delays = [1 2]
The resulting shift is 35.

New Delays = [1 3]
The resulting shift is 57.

New Delays = [2 3]
The resulting shift is 47.

New Delays = [1 4]
The resulting shift is 44.

New Delays = [2 4]
The resulting shift is 7.

New Delays = -1
```

The use of the GUI requires MATLAB version 6.5 or above. However, for lower versions of MATLAB, the program can be run by typing `prs_perturbation` in the MATLAB command window. The instructions are largely displayed in the command window and user inputs are mainly via the keyboard, with a small number of inputs entered through pop-up windows. An example is shown below for the generation of a QRB signal with $N = 11$, with the additional selection of a specification with no harmonics suppressed and number of signal levels $= 2$ from pop-up windows. The last line in the example terminates the routine.

```
Enter the length of the signal which can only be an odd
integer between 3 and 49999 inclusive.
11
Pseudo-random binary signal(s) of this length can be
based on the following classes:
Quadratic residue binary signal
Do you want to generate the signal? y/n [y]:y
The sequence is
    number level
       1     1
       2    -1
       3     1
       4     1
       5     1
       6    -1
       7    -1
       8    -1
       9     1
      10    -1
      11     1

PIPS = 99.585920%

The following PIPSE and EMINE values assume uniform harmonic
specifications, with the highest specified harmonic set to 5.
PIPSE = 86.375845%

EMINE = 79.916287%
Do you wish to save this data? y/n [y]:n
Do you wish to see any plots? y/n [y]:y
Mean value of the signal is 0.090909.
Press any key to continue except the Return key and some special
keys.

DFT magnitude at dc is 1.000000.
DFT magnitude at other harmonics within the period is 3.464102.
Press any key to continue except the Return key and some special
keys.

Enter a harmonic number to see its power spectrum
magnitude. If you do not want this function, enter -1.
Harmonic number = 3
The power is 0.154321.
Harmonic number = -1
Press any key to continue except the Return key and
some special keys.
```

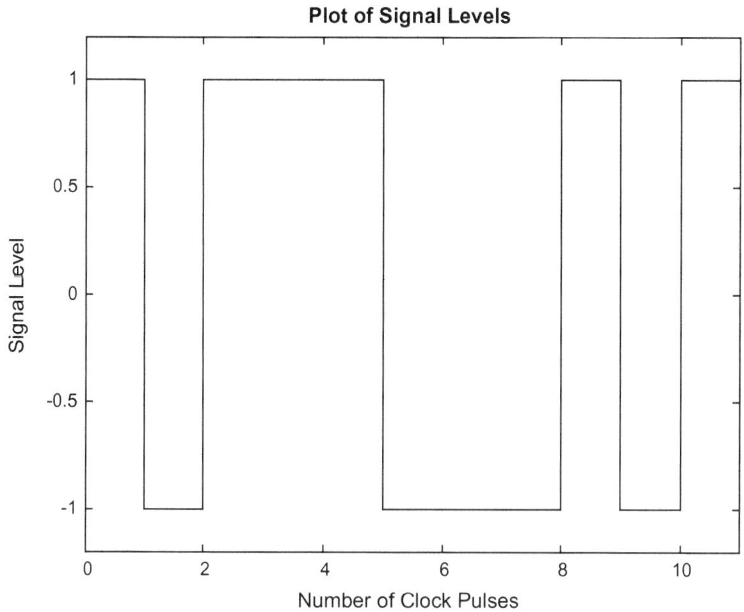

Fig. 8.2 Plot of signal amplitude for a QRB signal with $N = 11$ using *prs*

```
Autocorrelation value with no delay is 1.000000.
Autocorrelation value with 1 to 10 units of delay is -0.090909.
```

```
Do you want to generate a signal with a different
number of levels? y/n [y]:n
Do you want to generate a signal with a different
length? y/n [y]:n
Do you want a signal with a different harmonic
specification? y/n [y]:n
```

If the user chooses to save the data into a file, it will be saved into a file `perturbation_output` which can be opened using any text editor. If the user opts to view the plots, four plots appear which show the amplitude, DFT magnitude, power spectrum and autocorrelation function of the signal. These are illustrated in Figs. 8.2, 8.3, 8.4 and 8.5 for the QRB signal with $N = 11$.

The latest addition to *prs* is a subsidiary program which allows the user to generate direct synthesis ternary signals including suboptimal ones (see Sect. 2.4 for the related theory). To run the program, type `direct_synthesis` in the MATLAB command window. The instructions are displayed in the command window and user entry is through the keyboard. If the preferred period is not available, the program suggests the next available period, as shown in the example below with inputs from the user being underlined. If the user chooses to save the signal into a file, the program creates

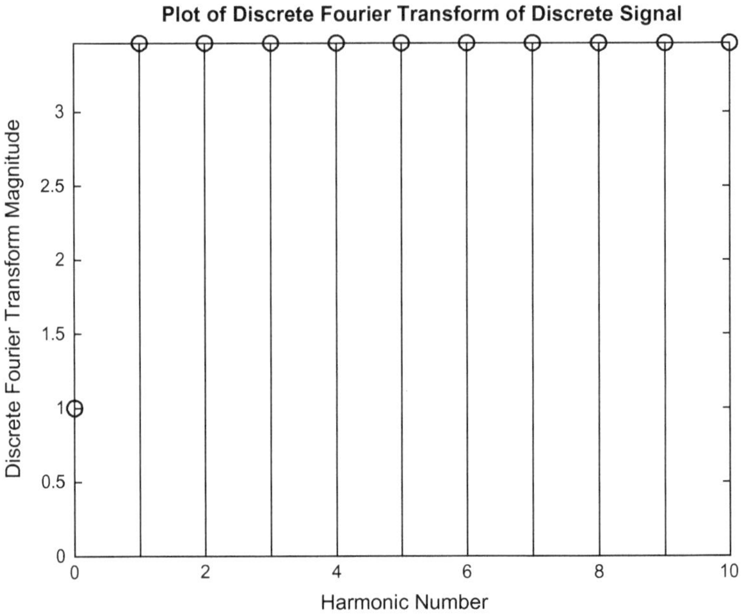

Fig. 8.3 Plot of DFT magnitude for a QRB signal with $N = 11$ using *prs*

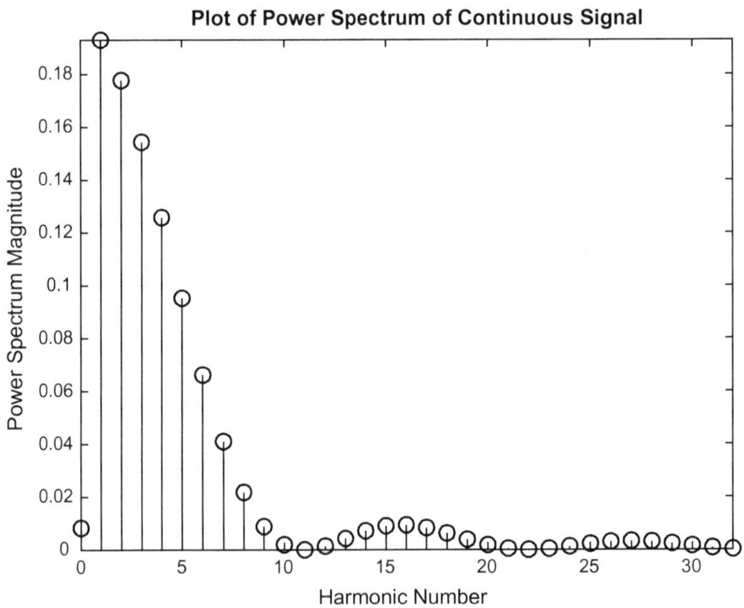

Fig. 8.4 Plot of power spectrum for a QRB signal with $N = 11$ using *prs*

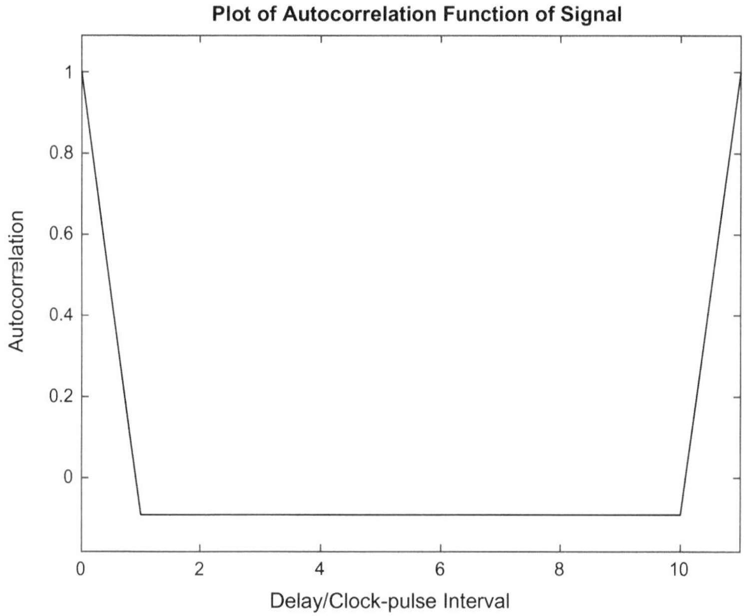

Fig. 8.5 Plot of autocorrelation function for a QRB signal with $N = 11$ using *prs*

a file ds_output which can be opened using any text editor. The program generates the plots of the signal amplitude, DFT magnitude and autocorrelation function if the user opts to view the plots. The last line in the example below terminates the program.

```
Welcome to the program for generating direct synthesis
perturbation signals.
Please enter the preferred signal period.
2500
No direct synthesis signal exists for the preferred
period.
The next period available is 2514.
Do you want to generate another signal? y/n [y]:y
Please enter the preferred signal period.
2514
More than one class of signals are available as basic
signals for implementing the direct synthesis approach.
Choose a signal class by entering the associated
character.
Quadratic residue binary signal : b
Quadratic residue ternary signal : q
b
```

```
The signal has been generated.
The root-mean-square value of the signal is 0.816497.
Do you wish to save the signal into a file? y/n [y]:y
Do you wish to see any plots? y/n [y]:n
Do you want to generate another signal? y/n [y]:n
```

8.2 GALOIS

The GALOIS program incorporates functions to generate PRML signals (see Sect. 2.3) for all prime and extension GF(q) for $2 \leq q \leq 128$. The program can be freely downloaded from https://sites.google.com/view/signaldesign/galois. It can display sum tables, product tables, inverses of all elements, reciprocals of all elements, all primitive elements and all primitive polynomials of degree n up to q^n of 16,384. User input is through a GUI. The program is based on the work described in Barker (1993).

Upon launching the GALOIS program, the user is faced with two initial choices, namely Sequences and Fields. If the user chooses Sequences, the program proceeds as below, with the entries at the same level of indentation indicating options for the user. For example, after entering the required field, the user may select View or Maximum Length Sequence.

- → Sequences
    - → Field (enter the required field)
        - → View
            - → Sum Table, Product Table, Elements (or all of them)
        - → Maximum Length Sequence
            - → Enter degree and select generating polynomial
                - → View (maximum length sequence)
                - → Pseudorandom Sequence
                    - → Design
                    - → Custom

The field elements in a maximum length sequence may be converted into signal levels to generate a PRML signal. Two options are available, namely, Design and Custom. A Custom option allows the user complete freedom to define these sequence-to-signal conversions. A Design option allows the user to implement predetermined conversions. In particular, for q = 3, 5, 7, 9, 11, 13, 17, 19, 23, 25, 27, 29 and 31, conversions are available for generating PRML signals with even harmonics suppressed and odd harmonics having uniform power. For q = 7, 13, 19, 25 and 31, conversions are available for generating PRML signals with harmonic multiples of two and three suppressed and all remaining harmonics having uniform power. For q =

31, conversions are available for generating PRML signals with harmonic multiples of two, three and five suppressed and all remaining harmonics having uniform power.

An example of generating a PRML signal with harmonic multiples of two and three suppressed from GF(7) with $N = 48$ is shown in Figs. 8.6 and 8.7. Figure 8.6 shows the selection of the field, degree of polynomial and primitive generating polynomial. The Design option, as well as the resulting signal, is shown in Fig. 8.7. The generated signal can be saved into .map or .dat files. Both file formats can be opened using any text editor. The .map format has additional information such as the generating polynomial and the sequence-to-signal conversion. The .dat file consists of only the signal values and is suitable for loading into MATLAB. For example, if the file is saved as signal.dat, then in order to load it into MATLAB, type `load signal.dat` in the MATLAB command window.

Truncated PRML signals (Sect. 2.3.2) are not included in the Design option. However, they can be generated using the Custom option in which case the user will need to perform the final truncation step.

If the user chooses Fields, the program proceeds as shown below.

→ Fields
    → Field (enter the required field)
        → View
            → Sum Table, Product Table, Elements (or all of them)
        → Primitive Polynomials
            → Enter degree of primitive polynomial
                → View primitive polynomials

Figure 8.8 shows the sum table, product table and elements of GF(5). In this case (for GF(5)), if the user opts to view the primitive polynomials, there are four primitive polynomials of degree $n = 2$, 20 primitive polynomials of degree $n = 3$, 48 primitive polynomials of degree $n = 4$, 280 primitive polynomials of degree $n = 5$ and 720 primitive polynomials of degree $n = 6$. Specifically, there are $\frac{\phi(q^n-1)}{n}$ primitive polynomials of degree n in GF(q), where $\phi(r)$ is Euler's totient function defined by the number of positive integers less than r and prime to r (Zierler 1959). The results can be saved into a .pol file which can be opened using any text editor.

8.3 Frequency Domain System Identification Toolbox

The Frequency Domain System Identification Toolbox (Kollár 1994) is a third party toolbox running on MATLAB. It has functions to cater for various steps in frequency domain system identification such as perturbation signal design, parameter estimation, model validation and uncertainty analysis. In terms of signal design, several classes of signals can be generated, for example, multisine signals, discrete inter-

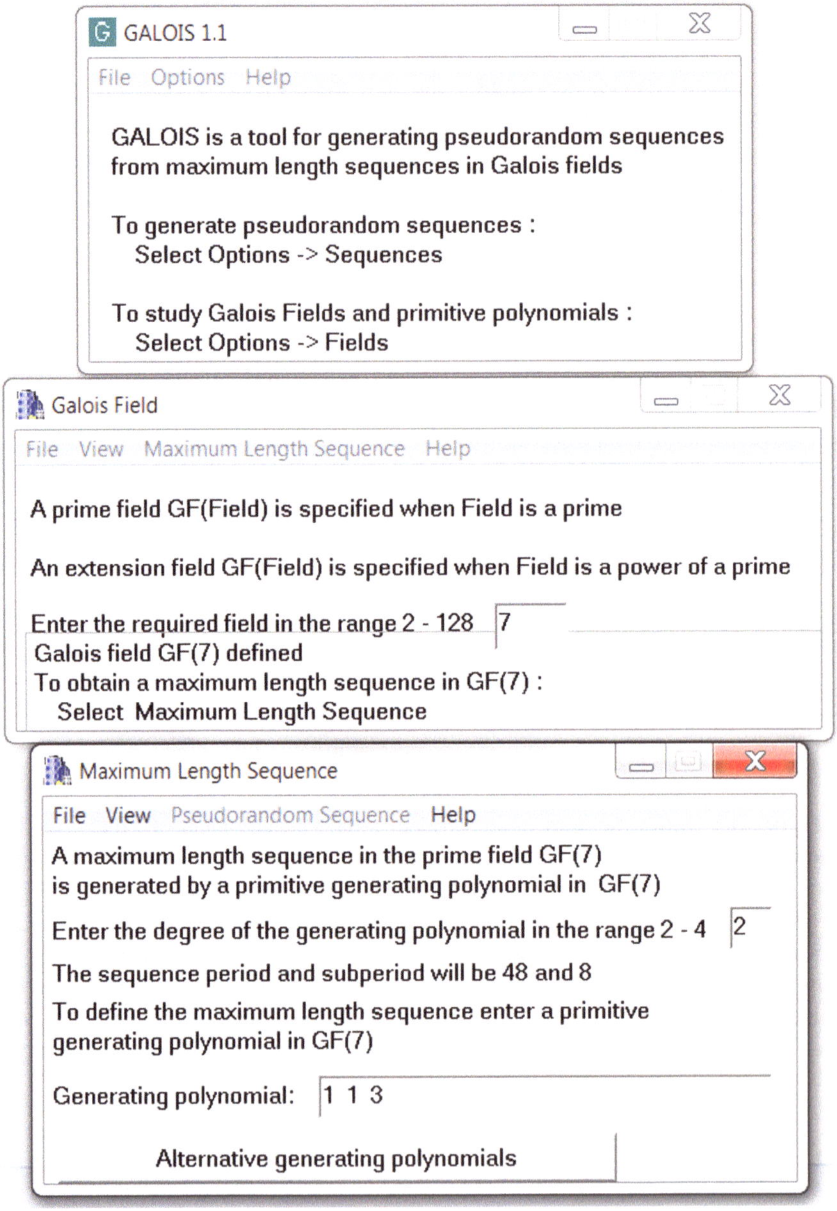

Fig. 8.6 Example of generating a PRML signal with $N = 48$ using GALOIS, showing the selection of the field, degree of polynomial and primitive generating polynomial

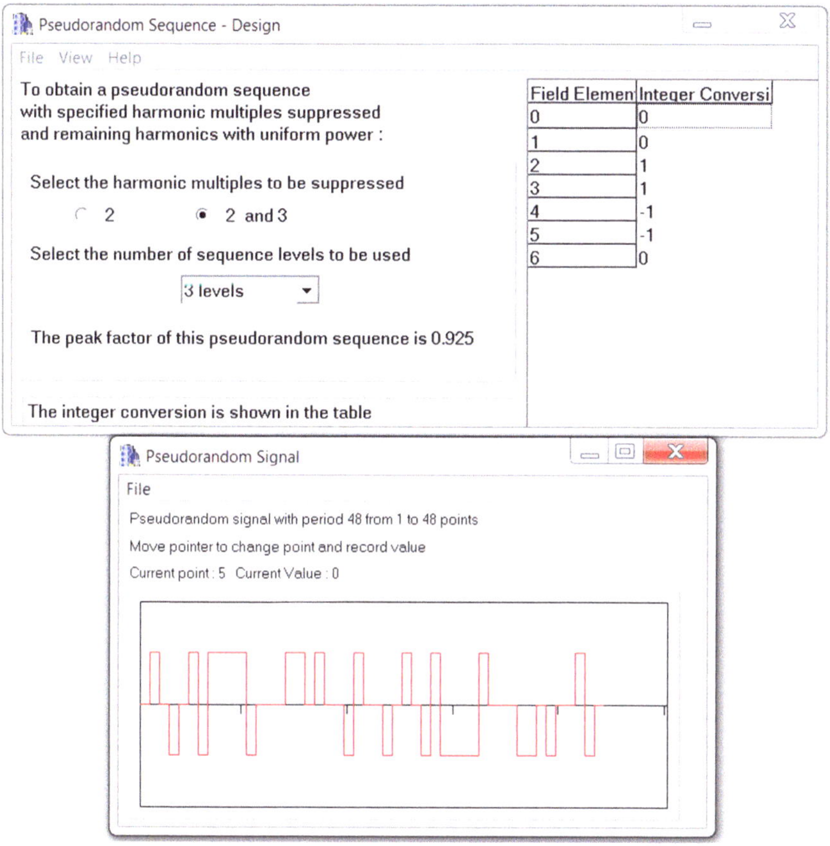

Fig. 8.7 Example of generating a PRML signal with $N = 48$ using GALOIS, showing the Design option as well as the resulting signal

val signals and optimal input signals. The reader may refer to Chaps. 2 and 7 of Schoukens et al. (2012) for further examples of signal design.

8.3.1 Generation of Multisine Signals

The time-frequency swapping algorithm described in Sect. 3.1 is implemented through the function `msinclip`. The function allows high flexibility in terms of user choices and comes with a relatively large number of input and output arguments. Explanation for these can be found through the `help` function. For most purposes, only a small number of arguments need to be specified. For example, for a specification with 50 consecutive harmonics (1, 2, 3, 4, …, 50), $N = 200$ with ZOH

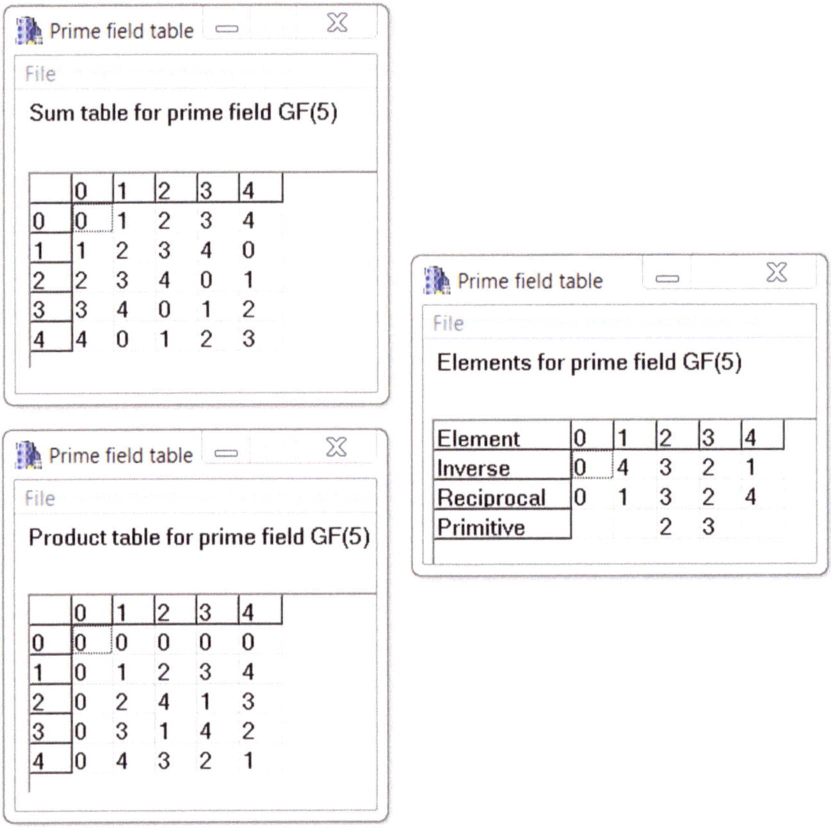

Fig. 8.8 Sum table, product table and elements of GF(5) using GALOIS

pre-compensation and Schroeder phases as starting phases, type the following code in the MATLAB command window to call the `msinclip` function:

```
[cx,crinfo]=msinclip(fiddata([],ones(50,1),[1:50]'/200),
struct('initset','Schroeder','crestmode','ZOH','N',200));
```

In the input argument, `fiddata` is a frequency domain data object. The first argument of `fiddata` is left unspecified using `[]` as this refers to the Fourier coefficients of the output of the system. The second argument of `fiddata`, entered as `ones(50,1)` in this case, is the required Fourier coefficients of the input signal corresponding to the frequency points specified in the third argument, `[1:50]'/200`. Some additional options for the design are specified using `struct`. The initial phases are specified to be Schroeder phases, by using `'initset','Schroeder'`. If the user chooses to start with random phases, this can be specified using `'initset','random'`. ZOH pre-compensation is specified using `'crestmode','ZOH'`. If a band-limited design is desired, this

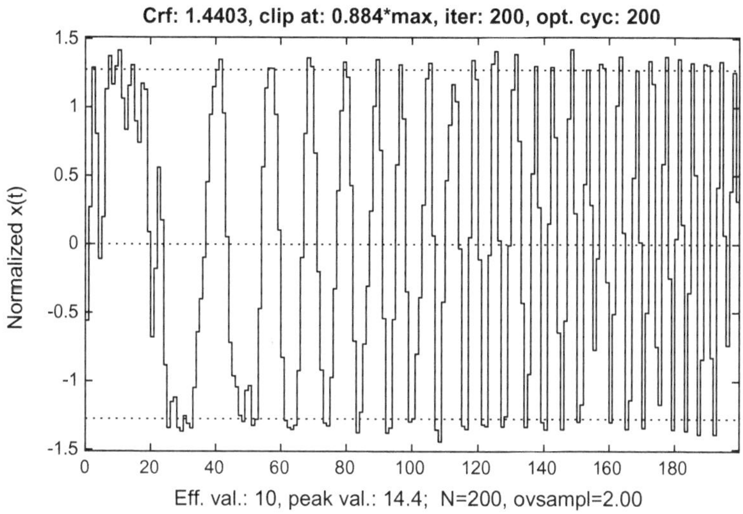

Fig. 8.9 Multisine design with $N = 200$ using `msinclip`

can be changed to `'crestmode','BL'`. The signal period is specified using `'N',200`. Additionally, the iteration number can be set to 500, for example, using `'itno',500`. The default is 200 iterations.

Typing the `msinclip` command given above leads to an optimisation being performed. The user can view the progress of the optimisation through a plot of the time domain signal. The end result is shown in Fig. 8.9. In the output argument, `cx` is the frequency domain data object of the designed multisine, where `cx.data` is a complex vector representing the Fourier coefficients of the generated multisine. For the case with ZOH pre-compensation, the Fourier coefficients are computed with the inverse of the ZOH transfer function being applied. The calculated coefficients are those of the ZOH-generated excitation signal. In this example, `cx.data` is a vector of length 50, and magnitude of 1 throughout, according to the given specification.

The output argument `crinfo` contains information on the crest factor. In particular, typing `crinfo.crx` gives the crest factor of the generated multisine which is 1.4403 in this case. Typing `crinfo.crxmax` gives the worst case crest factor of the multisine which is also 1.4403 for this example.

To generate the time domain signal corresponding to the Fourier coefficients, type the following in the MATLAB command window:

```
timeobject=msinprep(cx,200,1);x=timeobject.data;
```

The `msinprep` function generates a time series from the complex amplitudes of the multisine. The result is a time domain data object from which the time domain signal can be extracted.

Let us say 'snowing' (see Sect. 3.1) is to be added in harmonics 51–60. The `msinclip` command can be modified to

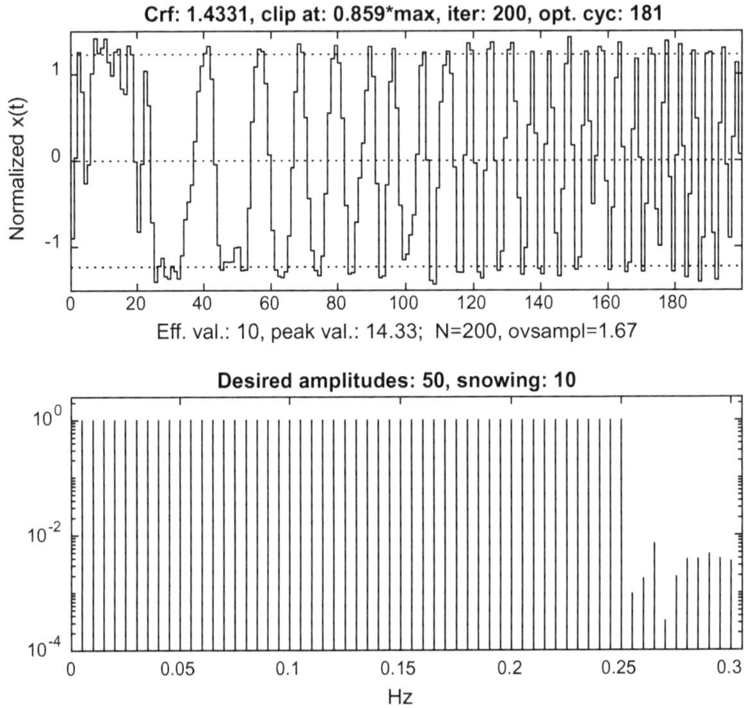

Fig. 8.10 Multisine design with $N = 200$ using `msinclip` with 'snowing'

```
[cx,crinfo]=msinclip(fiddata([],[ones(50,1);NaN*ones(10,1)],
[1:60]'/200),struct('initset','Schroeder','crestmode','ZOH',
'N',200));
```

In particular, the `NaN` vector specifies the 'snowing' input which corresponds to the last 10 frequency points (harmonics 51–60). A plot of the result is shown in Fig. 8.10, where it can be seen that the crest factor has improved slightly to 1.4331 due to the 'snowing'.

To generate multisine signals using the infinity norm algorithm, the `crestmin` function can be applied. The syntax is very similar to that for `msinclip`, except for some user options. In particular, `itmax` defines the maximum number of iterations for each value of p while `pmin` and `pmax` define the minimum and maximum values of p, respectively. Take for example a specification with 50 consecutive harmonics (1, 2, 3, 4, …, 50), $N = 200$ with ZOH pre-compensation. Assume that it is desired to set the maximum number of iterations to 100, and the minimum and maximum values of p to 2 and 128, respectively. Then, the following code can be used to call the `crestmin` function:

```
[cx,crinfo]=crestmin(fiddata([],ones(50,1),[1:50]'/200),
struct('initset','Schroeder','crestmode','ZOH','N',200,'itmax',
100,'pmin',2,'pmax',128));
```

The default setting is 50 iterations, minimum value of $p = 2$ and maximum value of $p = 256$.

8.3.2 Generation of Discrete Interval Signals

DIB signals can be generated using the dibs function. As an example, for a specification with 50 consecutive harmonics (1, 2, 3, 4, ..., 50), $N = 200$ with ZOH pre-compensation, the following code can be used:

```
[bitser,Puf,Ptot]=dibs(fiddata([],ones(50,1),
[1:50]'/200),200,1);
```

In the input argument, fiddata is defined as in Sect. 8.3.1. This is followed by defining the signal period as 200 and the sampling interval as 1. In the output argument, bitser is a time domain data object of the designed DIB signal, where bitser.data represents the time domain series. Puf is the useful power as a fraction of the total power and Ptot is the total desired signal power.

Further user options can be set in the input argument such as trialno which is the number of trials (each trial starts with a random-phase multisine design), the default being 25, and cyclemax which is the maximum number of iterations for a trial, the default being infinity. The setting of reconstr defines whether ZOH pre-compensation is applied. If this is entered as 'c', ZOH pre-compensation will be applied, whereas if it is entered as 'd', ZOH pre-compensation will not be applied. The default for this setting is 'c'. Additionally, a specification with even harmonics suppressed may be specified by declaring type as 'odd'.

To illustrate the design with only odd harmonics, take for example a specification with 25 consecutive odd harmonics (1, 3, 5, 7, ..., 49) with $N = 200$. Assume that no ZOH pre-compensation is required. The following code can be used:

```
[bitser,Puf,Ptot]=dibs(fiddata([],ones(25,1),
[1:2:49]'/200),200,1,struct('reconstr','d','type','odd'));
```

The resulting DIB signal is shown in Fig. 8.11. The useful power as a fraction of the total power is 87.7%.

DIT signals can be designed using dits. The syntax is the same as for dibs but with the possibility of specifying a design with harmonic multiples of two and three suppressed. This is done by defining type as 'oddnothird'. The signal period must be an integer multiple of six for the specification to be valid.

8.3.3 Generation of Optimal Input Signals

The function optexcit can be used to design perturbation signals with optimal power spectrum (see Sect. 3.5). For example, to obtain the design in Fig. 3.26, the

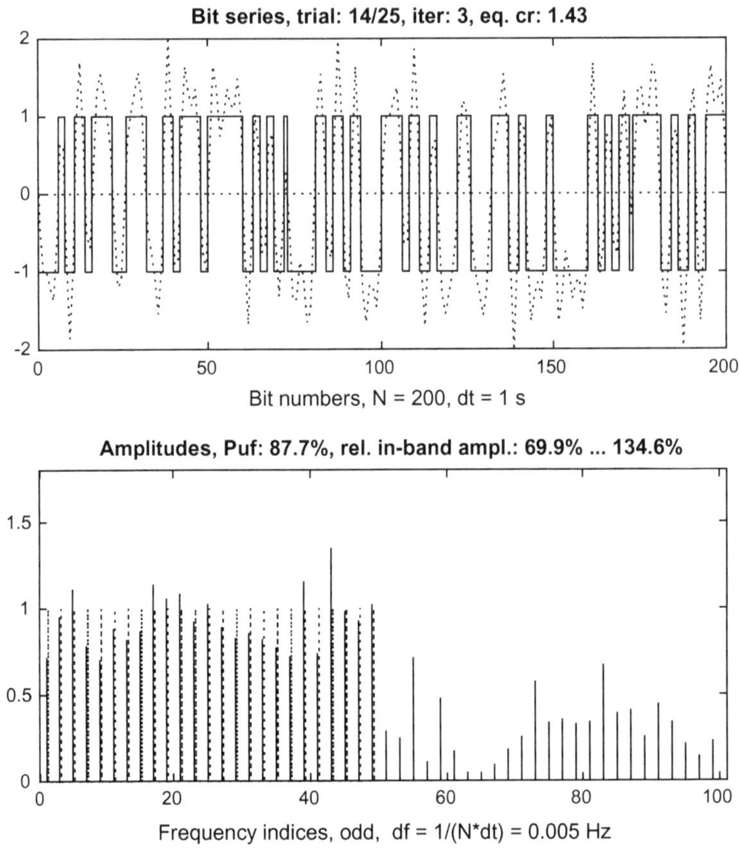

Fig. 8.11 DIB signal design with $N = 200$ using `dibs`

codes below can be used. Note that optimal input design requires knowledge (or at least a guess) of the system parameters. Hence, the system numerator and denominator need to be defined before calling the `optexcit` function. The vector `fixpar` specifies the indices of fixed parameters in the total parameter vector, defined in the form of `[num,denom,delay]'`, where the coefficients are in descending order in the s-domain, and in ascending order in the z-domain. In the example given, `fixpar=[6]` refers to fixing the delay since there are two parameters in the numerator and three parameters in the denominator. The sixth parameter is the delay.

The output plot generated from the codes below is shown in Fig. 8.12. With the optimal spectrum being found through `optexcit`, the spectrum can be easily realised using multisines.

```
%define system parameters
domain='s';
num=[3,2];denom=[1 5 10];
delay=0;
```

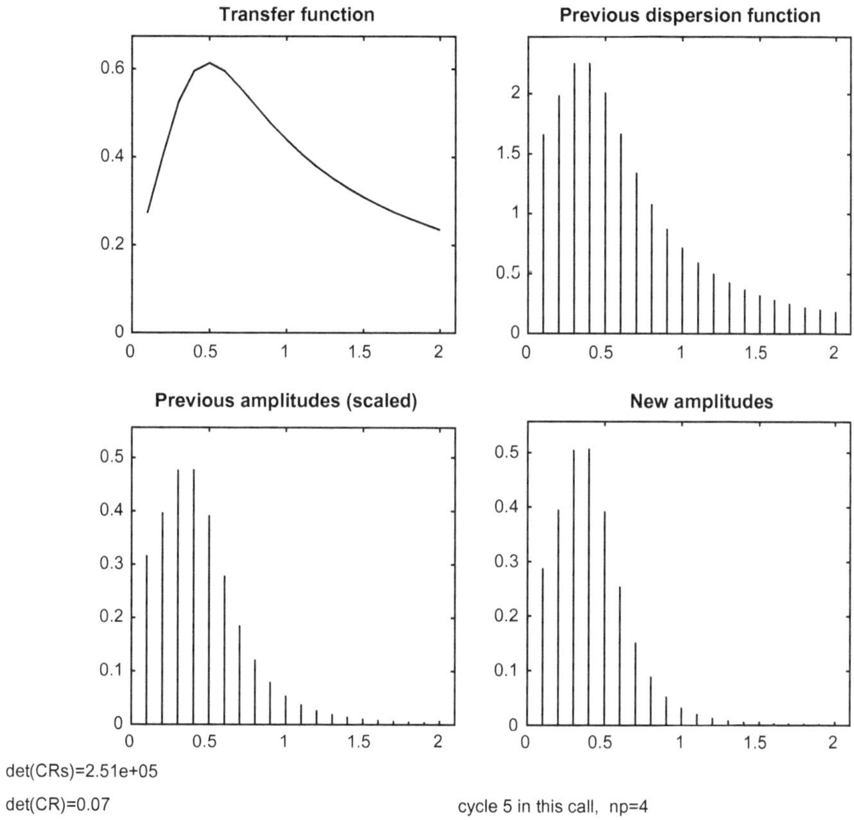

Fig. 8.12 Optimal input signal design using optexcit

```
fs=1;
%write parameters to a parameter vector
pdat=exppar(domain,num,denom,delay,fs);

%define the indices of fixed parameters in the
%total parameter vector
%fixing different parameters will give different result
%in this example, only the delay is fixed
fixpar=[6];
%define frequency vector
freqv=[1:20]'/10;
%define starting spectrum to be uniform
%total power is 1
Xstart=ones(length(freqv),1)/sqrt(length(freqv));
number_of_cycles=5;

%X is the generated amplitude spectrum corresponding to the
%frequency vector
%CR is the Cramer-Rao lower bound of the covariance matrix of
```

```
%the parameter estimates
[X,CR]=optexcit(pdat,freqv,[1,1],fixpar,Xstart,
number_of_cycles);
```

8.4 *multilev_new*

The *multilev_new* routine can be freely downloaded from https://sites.google.com/view/signaldesign/multilev_new. The routine runs on MATLAB. It incorporates functions to generate MLMH signals (see Sect. 3.3). The algorithm is based on the work by McCormack et al. (1995). PIPSE and EMINE are given equal weight in the later version of the algorithm (Tan and Godfrey 2004).

GUI is available for MATLAB version 6.5 or above. To bring up the GUI, type `multilev` in the MATLAB command window. In most cases, only four parameters need to be set, namely, the number of signal levels, harmonic multiples to be suppressed, number of excited harmonics and signal period. An example of a 3-level signal design with 10 excited harmonics, harmonic multiples of two and three suppressed and $N = 300$ is shown in Fig. 8.13.

For lower versions of MATLAB, the program can be run without GUI. For example, the equivalent design shown in Fig. 8.13 can be obtained by typing the following in the MATLAB command window:

```
Signal=multilev_new(300,[1 5 7 11 13 17 19 23 25 29],
ones(10,1),[2 3],3,1,100,[1:0.1:4]);
```

In the input argument, `300` is the signal period, `[1 5 7 11 13 17 19 23 25 29]` is the vector of excited harmonics, `ones(10,1)` represents the amplitude spectrum (uniform spectrum), `[2 3]` defines the harmonic multiples to be suppressed, `3` defines the number of signal levels, `1` selects ZOH pre-compensation (enter `0` if otherwise), `100` is the number of trials and `[1:0.1:4]` is the vector of allowable quantiser parameter values. In this example, since there are 31 allowable quantiser parameter values (1, 1.1, 1.2, 1.3, ..., 3.99, 4), the algorithm will make 31 trials with Schroeder phases as starting phases to select the optimum value. Then it will make another $100 - 31 = 69$ trials using the selected value with random phases as starting phases.

8.5 Input-Signal-Creator

Input-Signal-Creator is a suite of MATLAB programs which can be freely downloaded from https://sites.google.com/view/signaldesign/input-signal-creator. It incorporates functions to generate PRML signals. An important difference of this program compared with GALOIS is that the user does not need to know the basics of GF theory as the program automatically selects the required field based on the specifications. Truncated PRML signals are supported in Input-Signal-Creator. Two programs

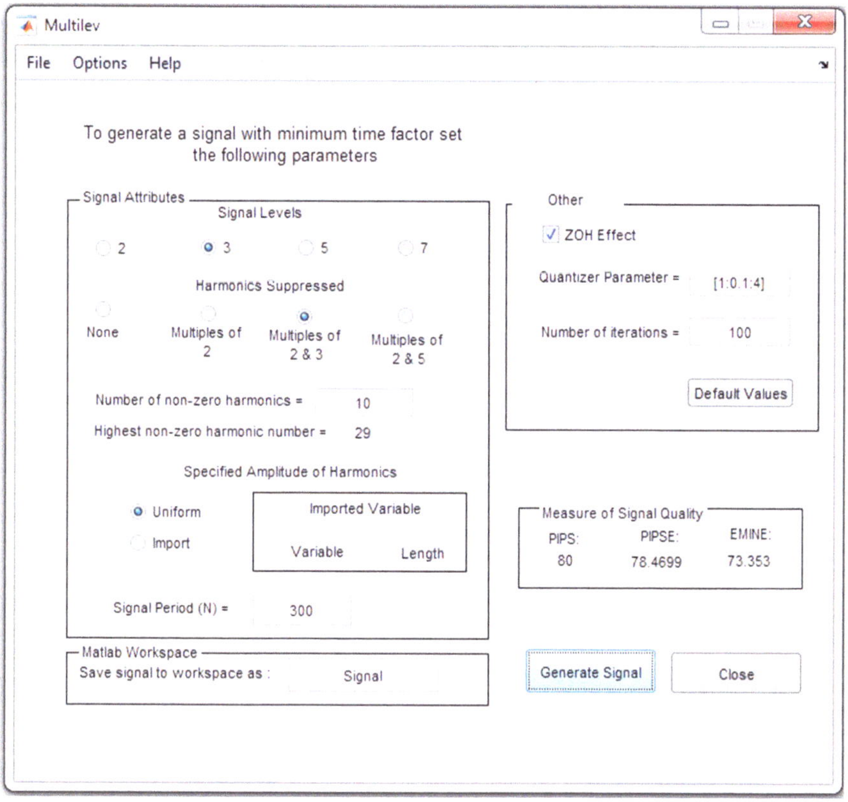

Fig. 8.13 Example of generating an MLMH signal with $N = 300$ using *multilev_new*

are available in Input-Signal-Creator. The first one is `createsignal` which generates a PRML signal given the specifications. The second one is `multiinputtool` which provides the periods of uncorrelated signal sets suitable for multi-input identification. Input-Signal-Creator is based on the work in Barker et al. (2014).

To run the first program, type `createsignal` in the MATLAB command window. The user will be prompted to enter the signal class (Class 0, Class 1 or Class 2; refer to the definition in Sect. 4.1.3), number of signal levels and signal period. The instructions are largely displayed in the command window, with a small number of inputs entered through pop-up windows.

As an example, assume that a 5-level signal with even harmonics suppressed and signal period close to 45 is required. Note that no knowledge of Galois theory is assumed. The MATLAB command window appears as follows, with inputs from the user being underlined. Some user entries, namely signal class (in this example, Class 1), number of signal levels (in this example, 5), performance measure to display (in this example, PIPS) and program termination, are achieved via selection of choices through pop-up windows. A very useful feature is that the program suggests the nearest periods where signals with the given class and number of levels are available.

```
Welcome to the program createsignal in the Input-Signal-Creation
Program Suite

The program creates input signals for system identification

The input signals are periodic with periodic power spectra and
autocorrelation functions

For multiinput system identification the program multiinputtool
is available for the determination of appropriate sets of
periods for the input signals

Every input signal has the following properties:

A period between 3 and 27384 inclusive
A power spectrum with equal power in all usable nonzero harmonics
A performance index for perturbation signals (PIPS) of at least
67%

Three classes of input signal, with two, three, five or seven
levels, may be created:

Class 0 signals where the power of all usable harmonics is
nonzero
Class 1 signals where the power of all harmonics that are
multiples of 2 is zero
Class 2 signals where the power of all harmonics that are
multiples of 2 and 3 is zero

Select the signal class from the menu

      The signal is Class 1

Select the number of signal levels from the menu

      The signal has 5 levels

527 Class 1, 5-level input signals are available with periods
from 24 to 27384


          Enter the preferred signal period 45


45 is not the period of a Class 1, 5-level input signal that
can be created
24 and 48 are the nearest periods of Class 1, 5-level input
signals that can be created


          Enter the preferred signal period 48
```

```
1 Class 1, 5-level input signal with period 48 has been created
The signal and its properties are available as the vectors
signal, spectrum, autocorrelation and PIPS
Select from the menu to display the signal and its properties
or to exit
 Performance Index PIPS 1

           81.01%

If the signal and its properties are satisfactory then the
program can be ended
The vectors signal, spectrum, autocorrelation, PIPS and LDC will
remain in the workspace for export, analysis or any other
purpose until another signal is created

Alternatively another signal can be created now

Select from the menu to create another signal or to end the
program

Program Ended
```

To run the second program, type `multiinputtool` in the MATLAB command window. The user will be prompted to enter the signal class (Class 1 or Class 2), the number of signal levels (if Class 1 was selected earlier) and the number of inputs through pop-up windows. The program then generates the sets of signal periods which meet the specifications. The signals in a set share a common period and are uncorrelated with one another in the set.

References

Barker HA (1993) Design of multi-level pseudo-random signals for system identification. In: Godfrey K (ed) Perturbation signals for system identification. Prentice Hall, Englewood Cliffs, NJ

Barker HA, Tan AH, Godfrey KR (2014) Object-oriented creation of input signals for system identification. IET Control Theory Appl 8:821–829

Kollár I (1994) Frequency domain system identification toolbox for use with MATLAB. The MathWorks Inc., Natick, MA

McCormack AS, Godfrey KR, Flower JO (1995) The design of multilevel multiharmonic signals for system identification. IEE Proc Control Theory Appl 142:247–252

Schoukens J, Pintelon R, Rolain Y (2012) Mastering system identification in 100 exercises. Wiley, Hoboken, NJ

Tan AH, Godfrey KR (2002) The generation of binary and near-binary pseudorandom signals: an overview. IEEE Trans Instrum Meas 51:583–588

Tan AH, Godfrey KR (2004) An improved routine for designing multi-level multi-harmonic signals. In: Proceedings of the UKACC international conference on control (Paper ID–027), Bath, UK, 6–9 Sept

Zierler N (1959) Linear recurring sequences. J Soc Ind Appl Math 7:31–48

Index

A. H. Tan and K. R. Godfrey, *Industrial Process Identification*, Advances in Industrial Control, https://doi.org/10.1007/978-3-030-03661-4

MIX
Papier aus verantwortungsvollen Quellen
Paper from responsible sources
FSC® C105338

If you have any concerns about our products,
you can contact us on
ProductSafety@springernature.com

In case Publisher is established outside the EU,
the EU authorized representative is:
Springer Nature Customer Service Center GmbH
Europaplatz 3, 69115 Heidelberg, Germany

Printed by Libri Plureos GmbH
in Hamburg, Germany